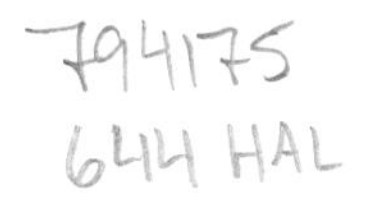

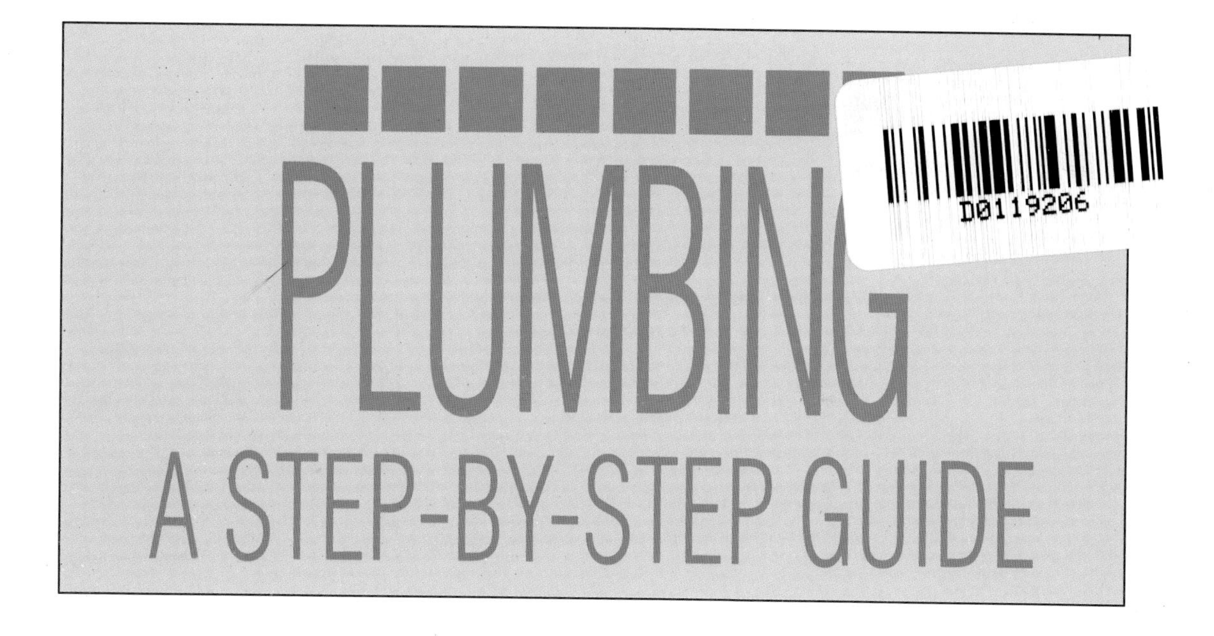

PLUMBING
A STEP-BY-STEP GUIDE

Photographs by Jon Bouchier, Simon Butcher, Simon Wheeler.

Illustrations by Kuo Kang Chen, Steve Cross, Paul Emra,
Pavel Kostal, Janos Marffy, Sebastian Quigley, Laurie Taylor,
Brian Watson, Andrew Green.

This edition published in 1993 by SMITHMARK Publishers Inc.,
16 East 32nd Street, New York, NY 10016.

SMITHMARK books are available for bulk purchase for sales
promotion and premium use. For details write or call the manager of
special sales, SMITHMARK Publishers Inc., 16 East 32nd Street, New
York, NY 10016; (212) 532-6600.

Produced by Dragon's World Ltd, 26 Warwick Way,
London SW1V 1RX, England.

Editor: Dorothea Hall
Designers: Bob Burroughs, Mel Raymond
Art Director: Dave Allen
Editorial Director: Pippa Rubinstein

ISBN 0-8317-4625-4

Printed in Italy

10 9 8 7 6 5 4 3 2 1

NOTICE
The authors, consultants and publishers have made every effort to make sure the information given in this book is reliable and accurate. They cannot accept liability for any damage, mishap or injury arising from the use or misuse of the information.

CONTENTS

INTRODUCTION

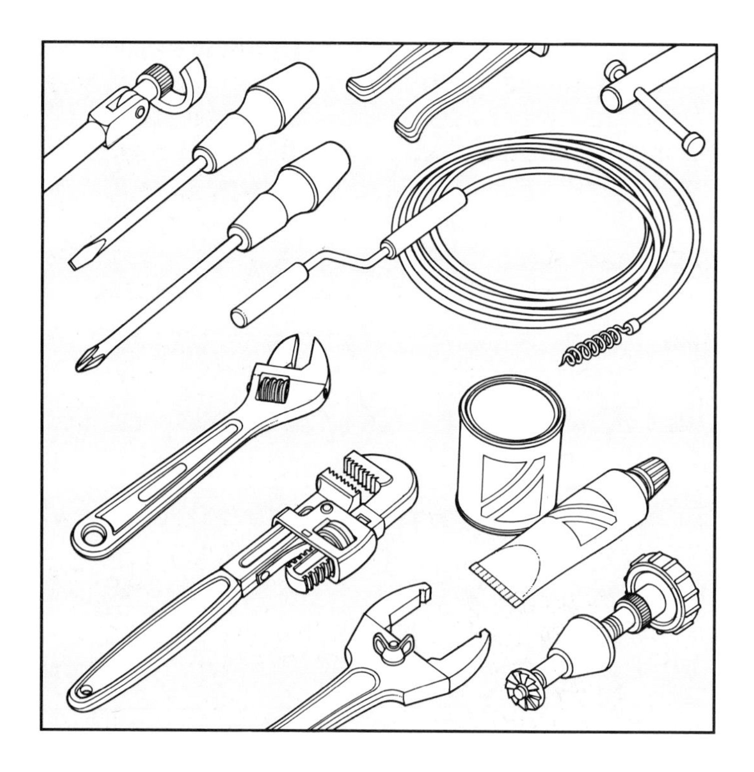

Many more people today are tackling their own home plumbing repairs and maintenance, not only because it is more convenient to cope with emergencies themselves, but because it is also much cheaper. Manufacturers have responded to this and have designed lightweight, easy-to-assemble fittings for almost every aspect of home plumbing repair and improvement. The following pages explain in detail everything the home plumber needs to know in order to do simple repairs both inside and outside the house; how to extend the plumbing system; install water filters and domestic appliances; add to the central heating system and replace the gutter system. With this professional advice, the home plumber is assured of achieving successful results every time.

Plumbing is work that many people, including do-it-yourself enthusiasts, are often wary of – the mere thought of the possible damage caused by gallons of water cascading through the house, should anything go wrong, is enough to give many people nightmares. But domestic plumbing systems are not as complicated as they may appear and modern plumbing materials and fittings are not difficult to use. Although you cannot hope to match the skills of a professional plumber, provided you approach each job with care, there is no reason why you should not handle most of the plumbing work around the home.

If you understand how the plumbing system operates, then you will be able to carry out the following repair and modernization techniques with absolute confidence.

In principle, the cold water for your domestic plumbing system comes, under pressure, directly from the public supply by a service pipe, or from a private well to a storage tank. Inside the house, the cold faucets will be supplied either directly from the main or indirectly from the cold water storage tank, see page 10. The hot water faucets are supplied through a separate pipe system which is invariably fed from the cold water supply. Your central heating will also be supplied by a separate pipe system.

While incoming water is supplied to the house under pressure, the waste system operates by force of gravity, see pages 11–12.

Dezincification

Depending on the type of soil through which the water runs before reaching your home; it may be hard or soft – either relatively high or low in minerals.

The type of water you have may also affect the type of materials that should be used for your plumbing. In areas where the water is extremely acidic, for example, brass fittings are prohibited

A centrally placed bathtub offers the maximum space for an all-round walkway – a useful feature in a shared family bathroom. As is the shower stall, fitted neatly into a corner of the room.

because of the process known as dezincification where the acidic water can, relatively quickly, dissolve the zinc from the brass (a copper and zinc alloy) and eventually destroy the fitting which, in turn, will cause leaks to occur.

In such districts, gun metal or solid copper fittings must be substituted for brass. In extreme cases, water may be so acidic it will dissolve lead or copper. Nowadays, plastic pipes and fittings are widely used, but as with any repair work and alterations, you should first check with your local building code.

Water softening

Where water is very hard, due to its high mineral salt content, hard scale is deposited on the inside of pipes and tanks, especially hot water cylinders and will even block the shower head. If the concentration is really high, scale can eventually block pipework and prevent heating elements from working efficiently – this drop in efficiency may be from 15 to 70 per cent. However, these effects can be greatly reduced by installing a water softener to the rising main at the point where it enters the house, see page 28.

Making changes

Apart from regular maintenance and essential repairs, there are many plumbing jobs that you might want to tackle yourself which could be a costly outlay if you were to employ a plumber.

The basic water supply/drainage systems in many houses are often inadequate for the needs of those who have to use them, and if the house is old, the fittings are likely to be ugly and showing the effects of age. In the bathroom, a ceramic basin may be chipped or stained, the enamel coating on the bath may be worn or chipped, and the toilet may have a noisy, inefficient tank. In the kitchen there may be a battered glazed earthenware sink or a utilitarian steel variety with chipped and stained enamel finish; neither lending itself to the installation of smart kitchen units. All can be replaced with modern stylish and efficient fittings in a wide range of colors to brighten up these purely functional rooms.

You may want to extend the systems to take account of a growing family or simply to make life easier. This may mean installing extra washing/bathing facilities, an extra toilet and, in the kitchen, making permanent connections for a washing machine, dishwasher and many other labor-saving devices that need a water supply. By exercising proper care and attention you will be able to tackle all these jobs without professional assistance.

Easily worked materials

Most modern plumbing systems are made with copper pipes with copper or brass fittings (although lead or galvanized steel pipes may be found in very old systems) and this material is easily cut, bent and joined, making alterations to the existing pipework a straightforward process.

Recent developments in plastic pipe and fittings have led to the availability of plastic hot and cold water supply systems. With their simple push-together or glued joints, these systems make plumbing jobs even easier. Plastic waste pipe systems have been in use for some years and are just as easy to assemble.

Obtaining the Necessary Permission

Depending on the type of work you intend carrying out, you may have to obtain permission before you start. Your local Building Inspector will be particularly concerned that any work you do does not lead to the water supply being contaminated, and he may not allow the use of some plumbing fixtures or fittings. Your Local Building Department will want to know that the work is carried out in accordance with the local Building Code, which lays down certain requirements for the way you arrange water supply and drainage from an appliance.

If you are simply replacing a fitting with a similar but more up-to-date item and you are not moving its position, it is most likely that permission will not be required. However, if you are using an unconventional fitting, or extending or modifying a drainage system, then you will almost certainly need approval. Some insurers will insist that at least part of the work – usually the plumbing and electrics – is carried out by professionals. If you are in any doubt at all about this, it is best to check with your local Building Code before starting work – it could save you a lot of wasted time and money.

Tools and equipment

In case of emergency, it is as well to keep to hand a few assorted faucet washers, O-rings, a length of hose and some hose clamps, a variety of nuts and bolts and metal washers. More specialized tools are illustrated opposite.

Slot and Philips screwdrivers are esential for fixing leaking faucets and other repairs.

Adjustable-end wrench has smooth jaws for fitting around small nuts and bolts, square and hexagonal fittings.

Pipe wrench (monkey wrench) has toothed jaws designed to grip pipe.

Spud wrench opens very wide and its toothless edges will fit around large nuts on sinks and toilets.

Basin wrench allows easy access to nuts located behind sinks and other similar places.
Rib-joint pliers open sufficiently wide to grip drain faucets.

Plunger used for dislodging clogged waste. Funnel-cup type is designed for toilets; the fold-flat type for drains.

Snake or drain and trap auger, stretches 10/20 feet to remove deep blockage in drain.

Closet auger is 3/6 feet long, has a crank handle and a protective housing to prevent scratching the bowl.

Valve seat wrench has one square end and one hexagonal end for removing worn valve seats.

Valve seat dresser smooths faulty nonreplaceable valve seats in old faucets.

Pipe joint compound is available in cans and tubes. Used for lubricating, protecting and sealing pipe threads at the assembly stage.

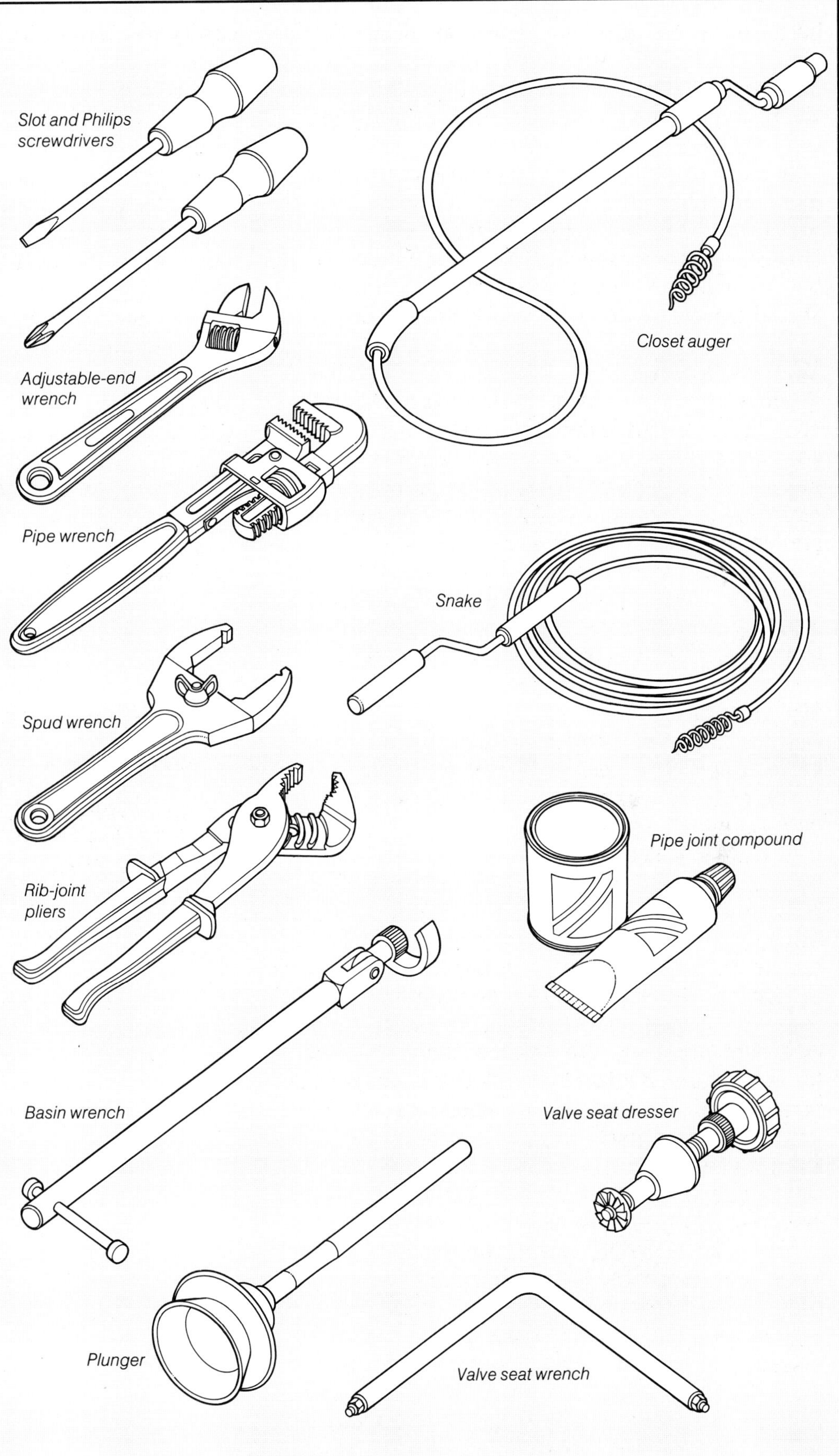

CHAPTER 1

THE PLUMBING SYSTEM

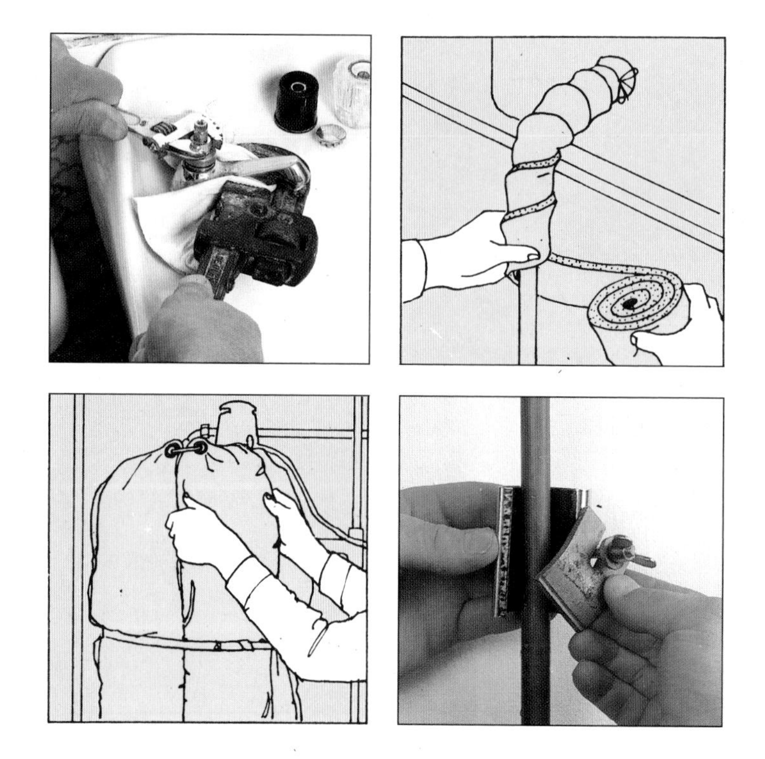

In order to carry out the simplest repair, you would first need to know how the plumbing system works in your home, where the shutoff valves are situated, and how the pipes are planned. This chapter explains how the system works and how to make simple repairs to faucets and toilet valve seats. It also explains how to unblock the waste outlets from the kitchen sink, toilet and drains — a relatively easy operation with the appropriate piece of equipment. Detailed information is also given for insulating pipes and cylinders, and for overhauling the central heating system before turning it on for the winter season.

HOW THE SYSTEM WORKS

Before you can attempt to do any repair work or modernization, you should familiarize yourself with the plumbing system in your home so that you can easily isolate the appropriate area, particularly in an emergency.

The direct system

In most properties today the direct supply system is installed where the main supply brings water (under pressure) to your house through a meter and a main shutoff valve. The supply pipes branch off to all inside water-using appliances and to outside faucets. The drain-waste system carries used water and waste away from the house by the main soil stack to the sewer or septic tank. The vent system gets rid of sewer gasses from each fixture and helps to maintain the correct pressure inside the drainpipes.

The indirect system

Most modern homes now plumbed with indirect systems are supplied by wells. However, in some remote and arid areas, outside water storage may still

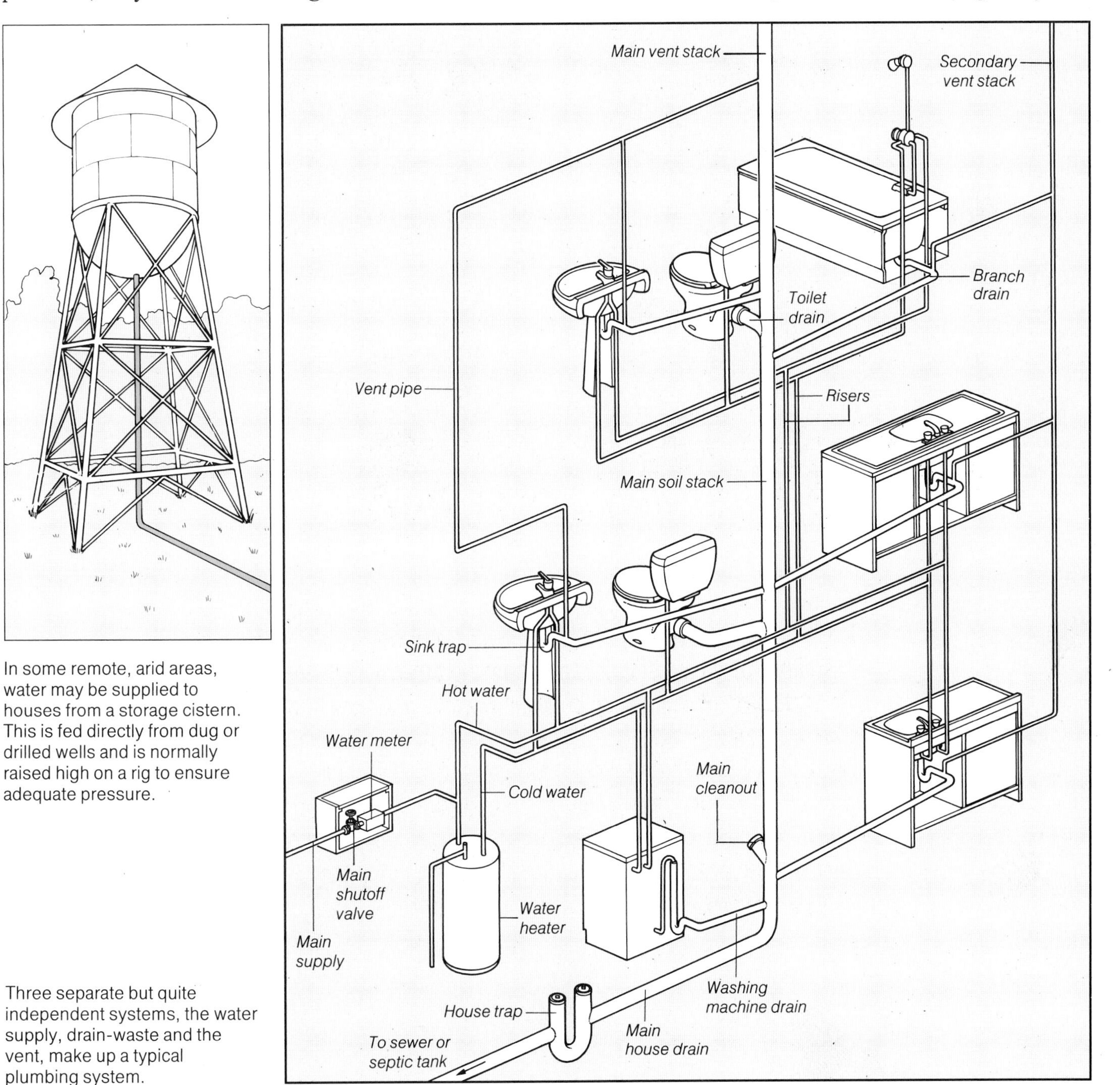

In some remote, arid areas, water may be supplied to houses from a storage cistern. This is fed directly from dug or drilled wells and is normally raised high on a rig to ensure adequate pressure.

Three separate but quite independent systems, the water supply, drain-waste and the vent, make up a typical plumbing system.

be found. The mains supply feeds directly into the storage which is normally raised on a rig, to guarantee adequate pressure. The hot water storage cylinder is also supplied with cold water from the same storage tank. Here the water is heated indirectly by immersion heater or the central heating system.

Modern houses are usually fitted with a single waste stack where all waste drains into a single soil pipe. Waste from the kitchen sink however, is discharged into its own trapped gully. Rain water is also carried away in a separate pipe.

A typical small bore hot water/central heating system. The two separate circuits supply the radiators and hot water faucets respectively. Hot water for the radiators is drawn from the top of the boiler and circulated with the aid of an electric pump. The hot faucets are supplied by a heat exchanger unit in the hot water tank.

HOT WATER/CENTRAL HEATING

In this particular combined hot water and central heating system water, supplied from a special feed tank, is heated by a boiler and then circulated around two separate circuits. One circuit connects all the radiators and the other sends hot water to a cylinder containing a heat exchanger which heats the water for the hot faucets. Most modern systems use narrow "small bore" pipes and have a pump fitted to force the water around the system.

The heat output is normally controlled by thermostats which automatically turn off the boiler when the preset temperature is reached and switch it on again when it drops below that level.

Always have your combined hot water and heating system overhauled every year by qualified engineers.

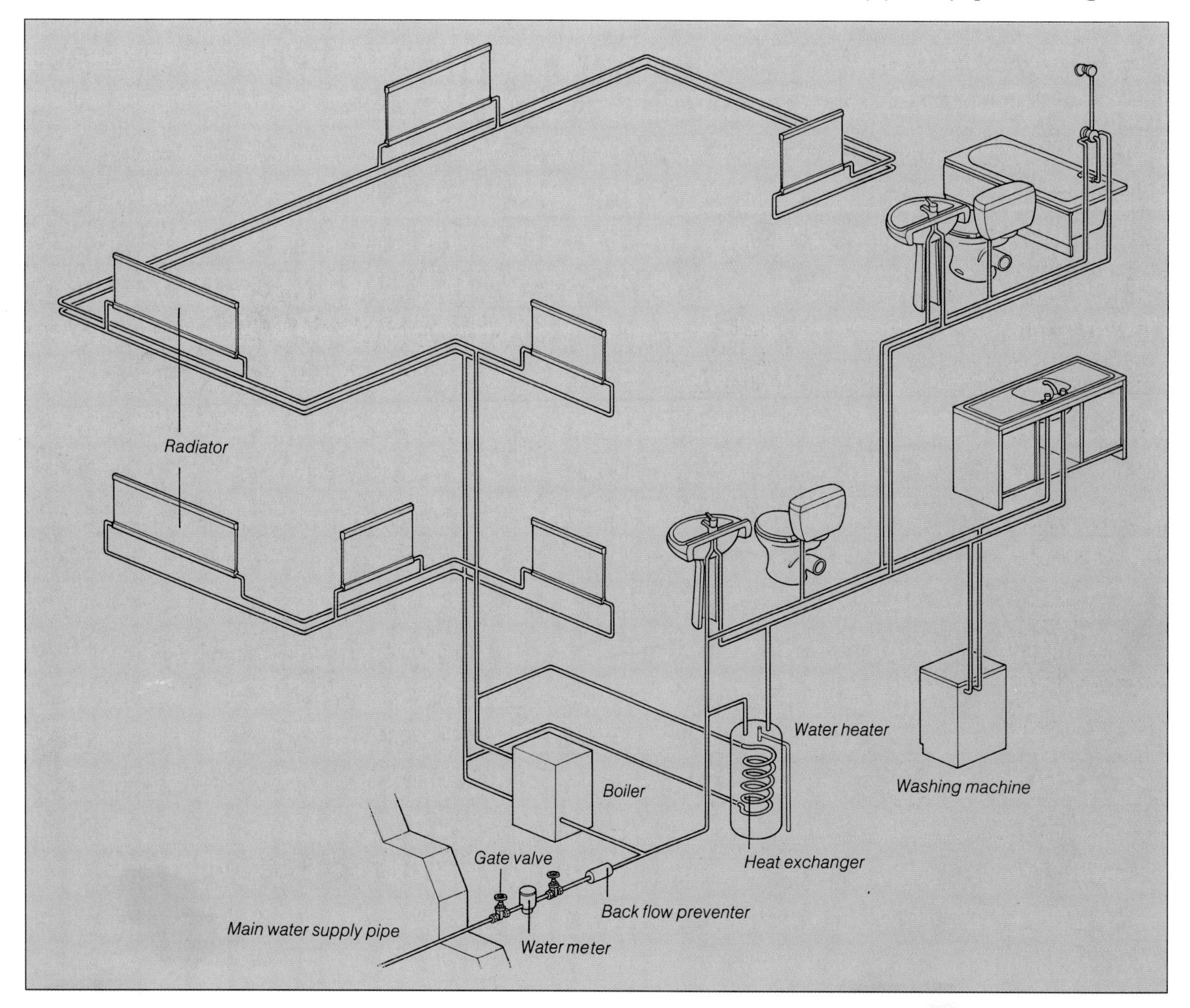

DRAINING THE SYSTEM

In order to carry out most repair work and all modifications or extensions to your plumbing system, it is necessary to drain part or all of the pipework. You should first familiarize yourself with the system so that you know where all the stop-valves and drain-cocks are situated and also which faucets and appliances are fed directly from the rising main, the cold water storage tank if any, and the hot water cylinder.

Draining parts of the system

It is not often that you will need to drain the system completely to do a particular job; it is usually only necessary to drain the affected circuit, keeping the rest of the system in commission and minimizing disruption.

Connections to the rising main or repairs to the various circuits within the house may mean turning off the main stop-valve fitted to the main supply pipe just after it enters the house. If you are lucky there will be a drain-cock immediately above the stop valve. Simply push a hose onto the outlet valve, lead it outside and open the valve to drain the pipe. If there is no drain-cock open the kitchen faucet until the water stops flowing.

When working on hot faucets, you can prevent too much hot water going to waste by shutting off the cold water main stop valve (which keeps the hot water cylinder topped up) first, then opening up the relevant hot faucet. This will keep most of the heated water still in the cylinder.

If the cylinder itself needs draining, turn off the immersion heater or boiler first. Close any stop-valve on the supply pipe from the cold water main and use the drain-cock which should be fitted at the base of the pipe to empty the cylinder. If there is no

•CHECKPOINT•

Adding extra valves

By adding extra gate valves to your plumbing installation, you will be dividing the system into relatively short pipe runs so that a particular area can easily be isolated without having to drain the whole system.

For this reason, it is best to install a gate valve on both cold and hot water supplies and on all major appliances.

It is also a good idea, when fitting new faucets, to fit small valves to the supply pipes just below the basin or sink. This will enable you to isolate the faucet easily and quickly when you need to repair it.

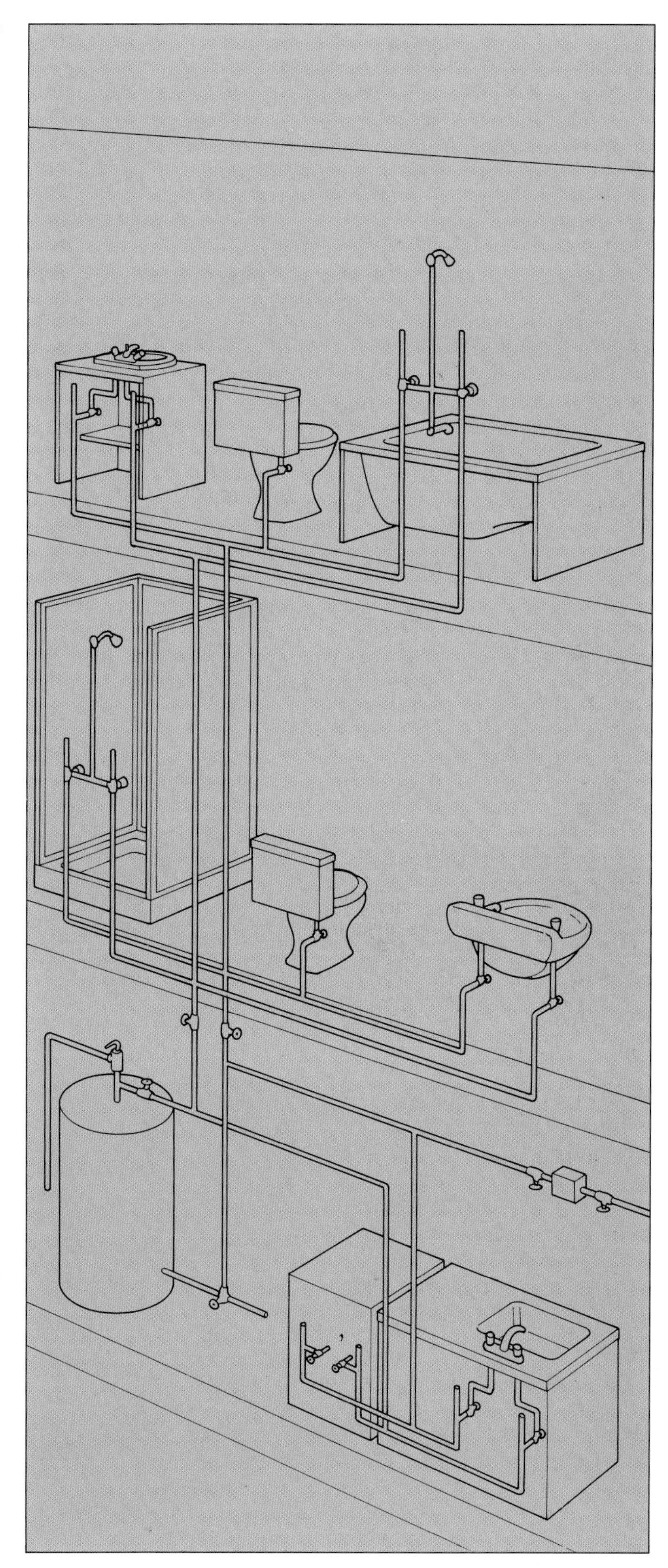

Water supply piping in a typical house. The supply pipe enters the house from the municipal supply or from a private well. A branch pipe feeds the hot water cylinder which is the beginning of the hot water supply system. Other branches feed to the individual fixtures and appliances.

drain-cock, the only way you can empty the cylinder is to disconnect the vent and the draw-off pipe at the top of the cylinder and carefully siphon out the water with a flexible hose, then connect the vent and draw-off all the water from the hot faucets to keep spillage to a minimum.

If a boiler circuit needs draining, turn off the boiler, tie up the float arm in the heating circuit feed and expansion tank in the basement and drain the circuit from the drain-cock next to or inside the boiler. If the boiler is part of a central heating system, this will need draining, too.

Draining the complete system

In an emergency you may need to drain the complete system. To do this, turn off the main stop-valve and open up all the faucets and drain-cocks. Once the trouble has been pinpointed, you can reinstate any unaffected circuits while you make repairs.

Refilling the system

In most cases all you need to do to refill the system is close any drain-cocks, open up any stop-valves or free any ball-valve float arms and close the faucets when water flows from them.

• CHECKPOINT •

Correcting a dripping faucet

Faucets which can not be turned off fully or which drip constantly need rewashering.

Turn off the water supply and open the faucet to drain the pipework. Depending on the type of faucet, either unscrew the protective cover and lift it up, or remove the shrouded handle by removing the retaining screw – under the indicator button in the top or at the side, or can the handle be prised off?

Hold the faucet spout to stop it turning and unscrew the head gear nut. Lift out the head gear and unscrew the nut holding the rubber washer underneath. Discard the washer and fit a new one. Then re-assemble.

Removing the headgear of a bathroom faucet after unscrewing the cover; grip the spout to prevent it from turning.

Fitting a replacement washer to the headgear after unscrewing the retaining nut to remove the old one.

• CHECKPOINT •

Correcting a dripping overflow

If an overflow pipe drips constantly, it indicates a problem with the ball-valve in the tank concerned. First check the float for leaks or bend the float arm so that it closes the valve fully. If this does not work, rewasher the valve.

Turn off the water and drain the pipework. Disconnect the float arm and unscrew the cap from the end of the valve. Push out the piston.

Hold the piston body with pliers and unscrew the cap from the end. Prise out the rubber washer and fit a new one.

Re-assemble the piston, lubricating it with petroleum jelly. Then re-assemble the valve and float arm before restoring the water supply.

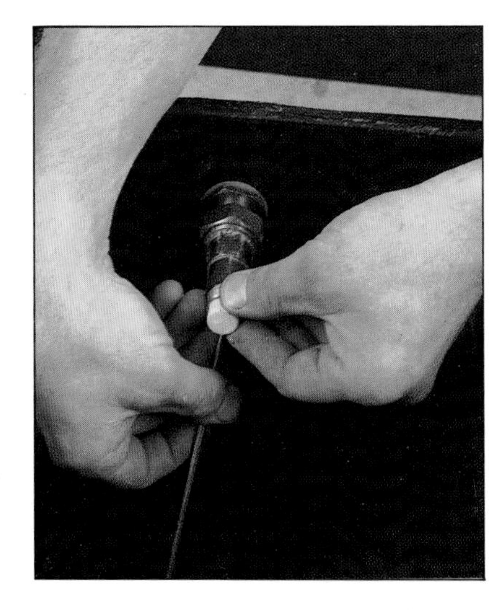

Prising the piston from a ball valve, using a screwdriver in the slot beneath, after removing the float and piston cap.

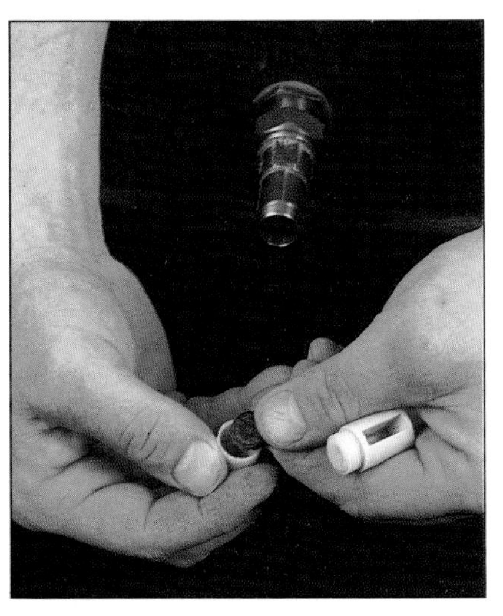

Inserting a new washer into the piston after dismantling the piston by unscrewing the two halves.

If you have drained a wet central heating system, you will have to open the radiator bleed valves (see page 20) to allow air to escape as they fill with water, closing them when water appears.

Curing an airlock

Air trapped in the system can cause a faucet to jump and splutter or fail completely. Air locks can be overcome simply by connecting a hose between the kitchen cold faucet and the faucet of the affected circuit. Open the latter and then the former, when the pressure of the water main will force the air bubble out of the system. If necessary, repeat the process until the water runs freely.

The low-level WC cistern

One of the most common plumbing problems is the WC cistern that will not operate properly. The cistern may fail to empty efficiently, or it will not refill correctly: the water either flows in too slowly or the cistern is constantly overfilling, see page 14. When the flushing arm is operated, it activates a siphonic action that empties the water from the cistern down the flush pipe and through the bowl. The usual cause of failure to flush properly is a worn valve seat, which fails to seal the mouth of the siphon efficiently when the metal plate is raised by the flushing arm. Water escapes around the edges of the valve seat instead of being carried upwards to start the siphonic action.

Replacing a WC valve seat

Before you can tackle a toilet repair, you should familiarize yourself with the individual parts and establish whether you are dealing with a wall hung (1) or bowl mounted tank (*right*). Flush tank empty.

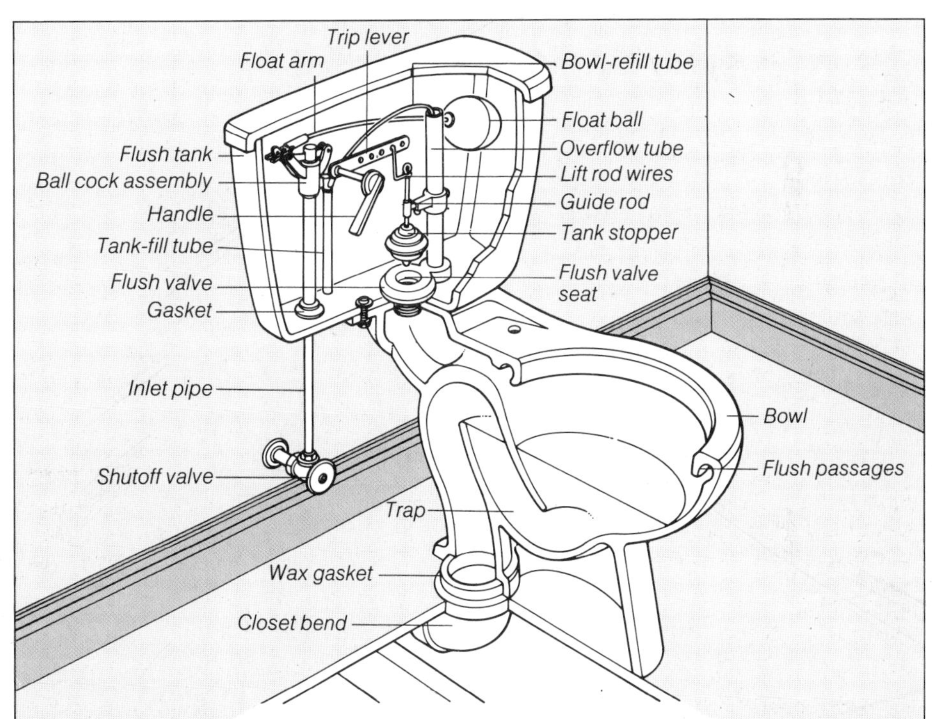

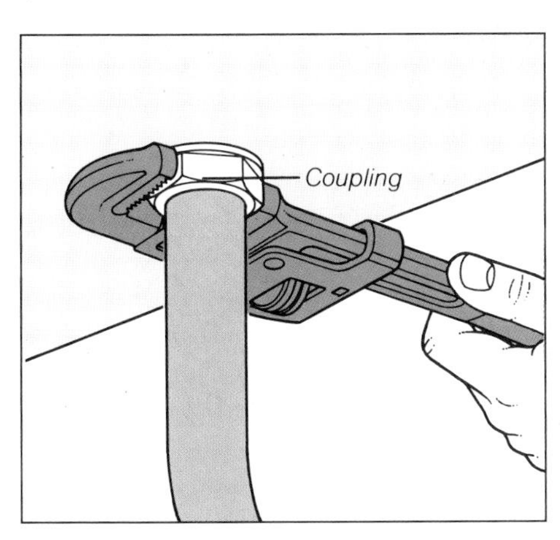

1 Remove old stopper, guide rod, lift the wires or chain. Loosen coupling under tank and remove pipe. Unscrew locknut; remove valve seat and gasket.

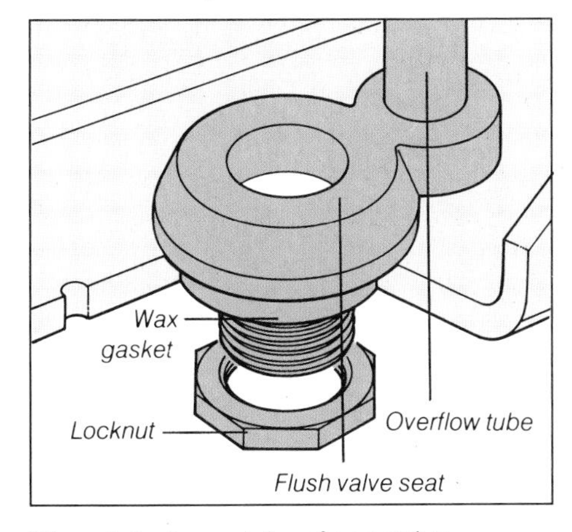

2 Insert discharge tube of new valve assembly through tank bottom. Place overflow tube towards the ball cock and tighten locknut.

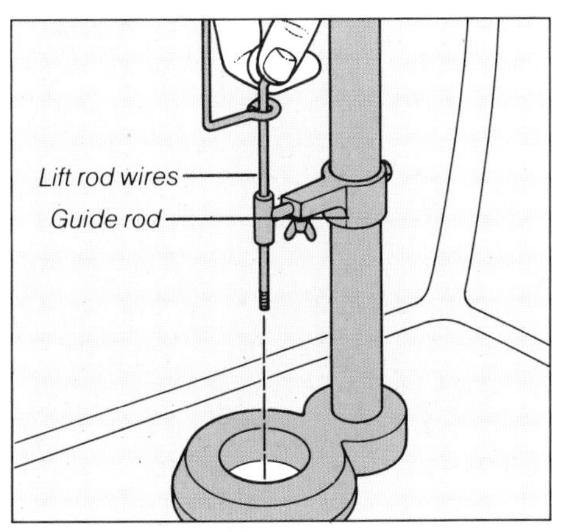

3 Centre the guide rod on the overflow tube and install the lift wires above the valve seat.

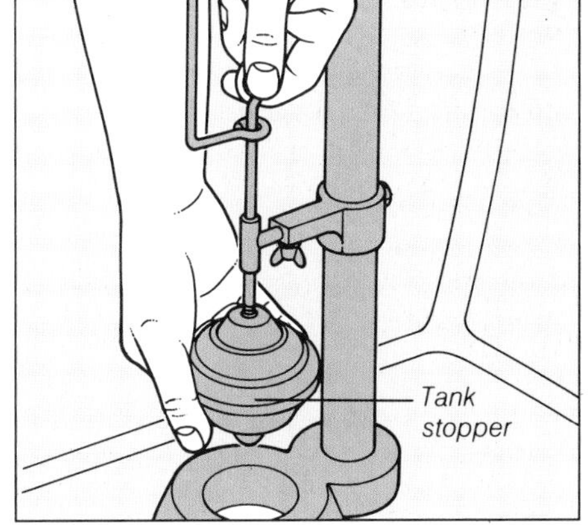

4 Screw the stopper onto the lower lift wire, aligning it with the centre of the valve seat.

FREEING BLOCKED WASTE OUTLETS

It is much better to treat a sluggish waste outlet before it becomes completely blocked and a real emergency.

Scalding water is particularly effective against a build-up of grease, at a kitchen sink, but the blockage may also be caused by a small object, such as a small utensil, hairpin or button which lodges itself across the pipe and other small particles of waste become trapped and very quickly build up into a fairly solid blockage.

An ordinary rubber plunger and drain auger are the most effective tools to use for freeing a blocked waste outlet in hand basins and sinks.

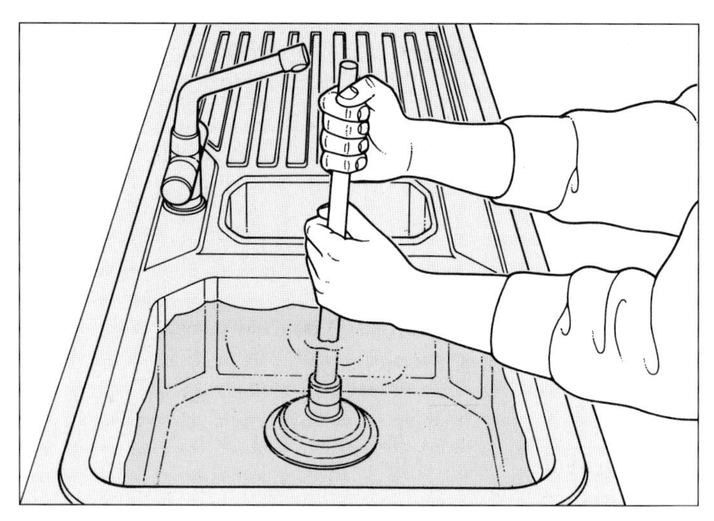

Pumping the plunger rapidly up and down for about 10 strokes; then quickly jerk the plunger away to remove the blockage.

Using a plunger

Begin by plugging the overflow opening with soft cloth and allowing enough water in the basin to cover the plunger cup. Coat the rim of the cup with petroleum jelly and place in centrally over the plughole. Without breaking the seal between the sink and the cup, pump vigorously up and down. Then quickly release the plunger when the blockage should be free. If not, then use the drain auger, or try cleaning the trap.

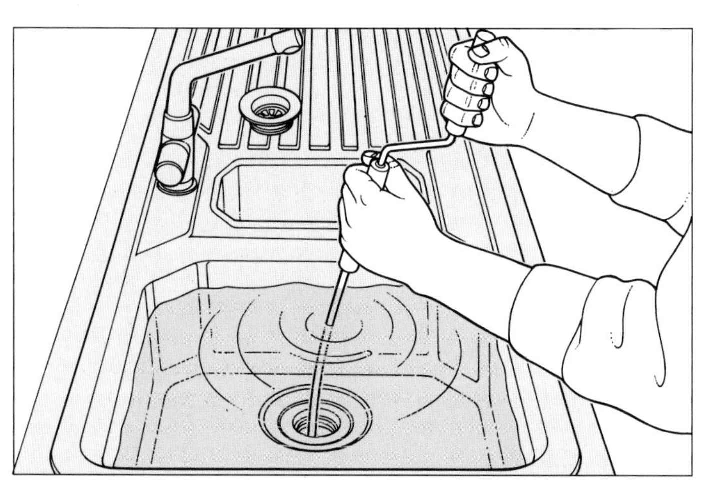

Feeding the auger deep into the waste pipe; alternately loosen and tighten the thumbscrew on the handle. Hook into the blockage, slowly move the auger backwards and forwards and winding in the same direction.

Using a drain auger

Should the plunger not work, then try the drain auger (snake). This is a very flexible metal coil that can be fed through the pipes until it reaches the blockage. The drain auger can be inserted into the basin if the strainer can be removed from the plug recess. Feed the auger into the waste by winding the handle clockwise. When you reach the blockage, move the auger backwards and forwards slowly while still winding, then slowly withdraw it, winding in the same direction.

Clearing the trap

Most modern sinks and basins have a removable bottle trap that forms part of the U-bend, and is situated directly below the plug outlet. Place a container beneath the U-bend to catch the water in the trap. Unscrew the bottom half and probe (using a small piece of wood) inside and around the inner pipe where waste collects. Clean and replace the trap. Turn on the water to check that it is now free.

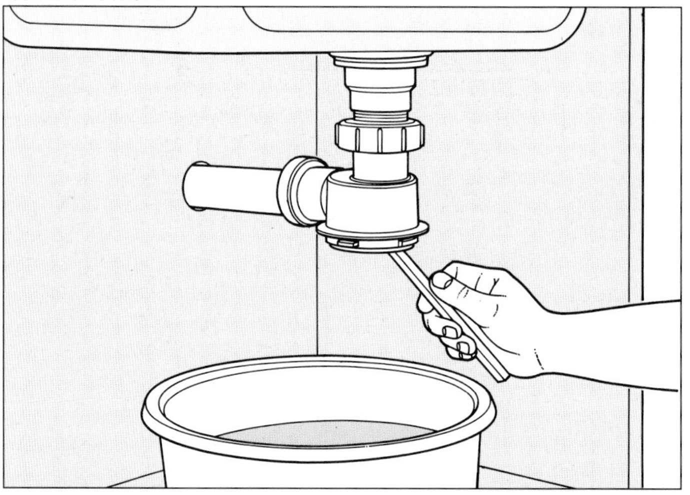

Unscrew the lower half of the bottle trap over a container and, using a small length of wood, probe inside to release the blockage. Clean the trap and reassemble.

Clearing the rainwater gully

An outside gully usually gets blocked with leaves and grit blown by the wind. You can use a hose under mains pressure where the force of the water may unblock it, or you could first try cleaning the edges of the grid (to keep a good fit) and then use a trowel to scoop out leaves and grit that may be the course of the blockage. If not, then use a drain auger or drain clearing rods.

Clearing the main drain

For bigger drains and WCs, special drain augers and clearing rods can be hired for the day.

The main cleanout is usually situated near the bottom of the soil stack where the main drain leaves the house.

Begin by placing a large pail underneath the cleanout to catch the flood of waste water in the drainpipe. Using a pipe wrench to remove the plug, open it slowly to control the flow of waste. Use either a snake, hose or hose with a balloon bag to remove the obstruction; then flush well with water. Apply pipe joint compound to the plug and recap the cleanout. If none of these methods is successful, then try cleaning the house trap.

Before opening the house trap, spread newspapers to catch the overflow of waste. Loosen the plug nearest to the outside sewer line. Using a snake, probe the trap and its connecting pipes. When the water begins to flow, cap the trap, wait for it to subside and then open up both ends. Clean out with a wire brush. Recap and flush the pipes through.

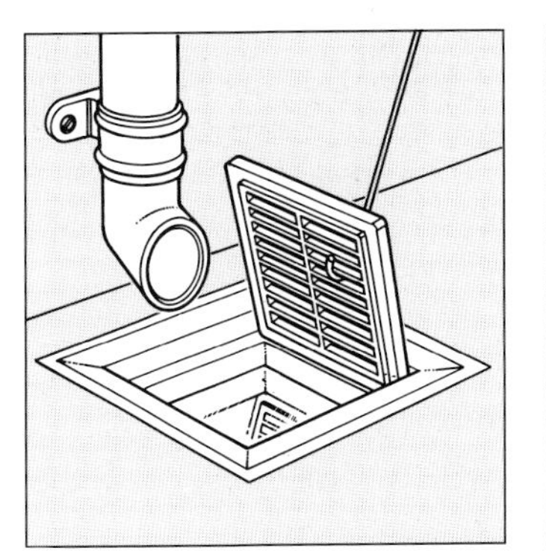

Clean out the recessed edges of the gully so that the cover will fit snugly.

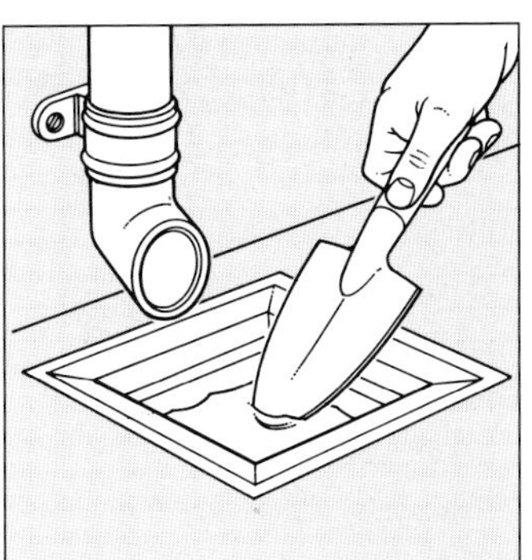

Regularly scoop out the collection of grit and leaves to keep the gully free.

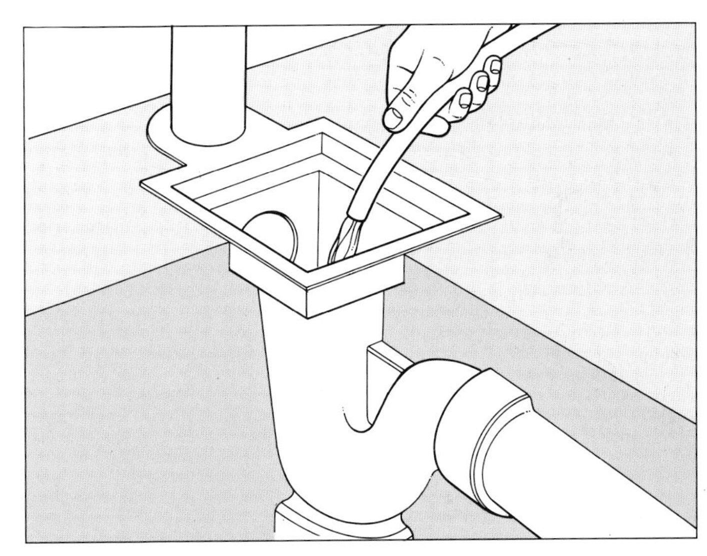

Above: Using a hose under mains pressure to free a blocked gully.

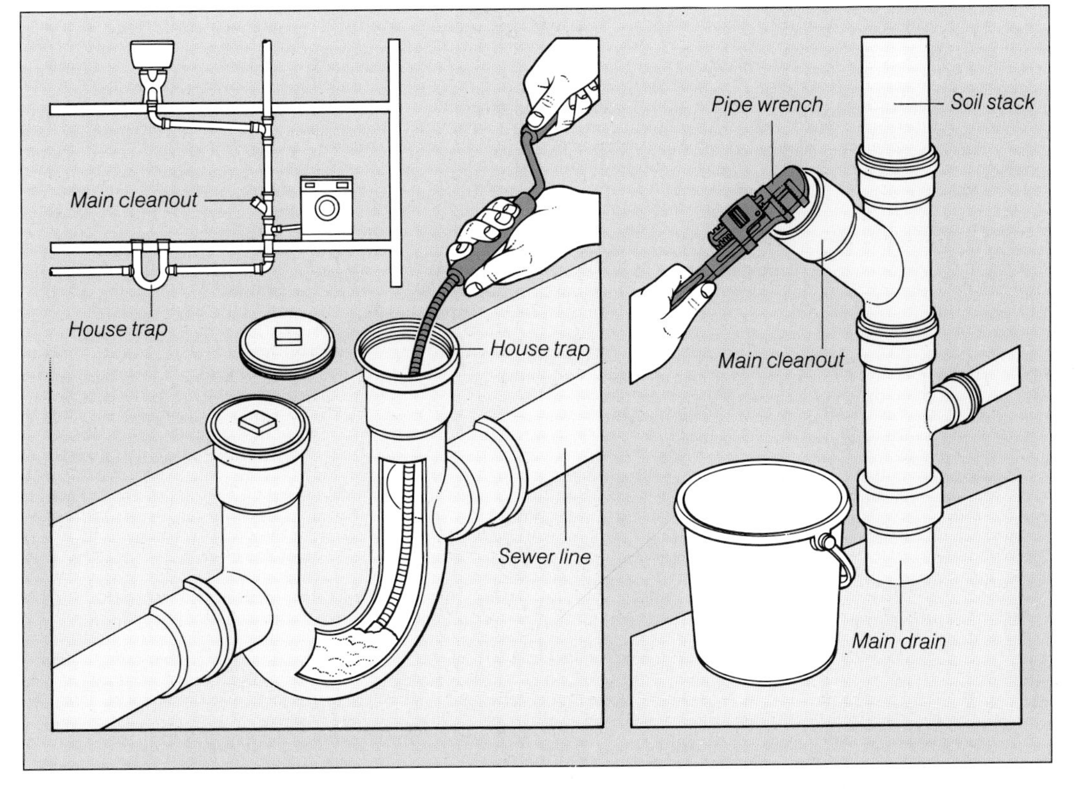

Far left: The main cleanout is usually positioned at the bottom of the soil stack, where waste enters the main drain.
Centre: When snaking through a house trap, work slowly allowing the water to drain gradually
Left: Using a pipe wrench to open the main cleanout, having first placed a large pail underneath to catch the water overflow.

INSULATING PIPES AND CYLINDERS

Of course, it is not just the heat loss from the house that can cost you money. There is also the heat loss from your plumbing system to be considered.

Hot water savings

The hot water system is the most obvious candidate for insulation, and you should start by fitting the hot water storage tank with as thick a jacket as possible. Do not skimp in the belief that too much insulation will ruin the effectiveness of your airing cupboard. No matter how much insulation you add, enough heat will still escape from the tank to meet your needs.

Insulating the hot water pipes as they run through the house is a rather less attractive proposition, for the simple reason that it can be expensive to do, yet yields little in the way of tangible savings of heat. For this reason, the best advice is to insulate only selected hot water pipes. Choose those that run in the attic or against outside walls, for example, plus any really long runs elsewhere in the house.

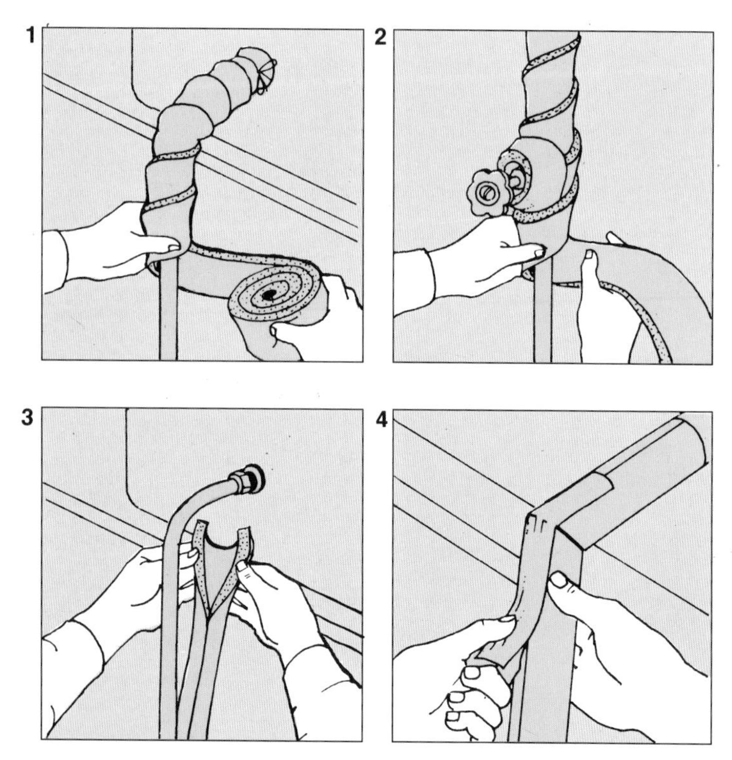

1 Starting at the tank, wrapping overlapping turns of insulating bandage round the pipe; tie the insulation at the start.
2 Wrapping the bandage round the body and neck of a valve; pay special attention to these as they must operate in an emergency.
3 Slipping split plastic foam tubing over a pipe; choose the correct size for the pipework and butt-join lengths before taping together.
4 Insulating a right-angle bend; cut the ends of the insulation to 45° miter with a sharp knife and tape together.

Cold water insulation

The idea that heat lost from the cold water system can cost you money may strike you as odd at first. But think about it. Is it not this that causes pipes to freeze and burst in winter? So to avoid the misery of frozen pipes it is absolutely vital to thoroughly protect every bit of pipework that is at risk.

That means every single pipe in the roof or attic, including overflows and vent pipes from the central heating and hot water systems. The cold water storage tank, and feed and expansion tank will also need protection, and do remember that, even if tanks and pipework already have some protection, if you intend to improve the insulation in the roof, existing insulation may need to be increased. After insulation, the roof will be a good deal colder, and that increases considerably the risk of a freeze-up.

And what about pipework elsewhere in the home? Any pipes to garden hose bibs and so on obviously need rather more insulation than usual once outside the house, and there is a slight risk of freezing in pipes run against an exterior wall, so these should be protected too.

Insulating pipework

Having decided which pipes need insulating, how do you set about it? Well, there are two main types of insulation available. Perhaps the easiest to use is plastic foam tubing, already split down the side so it can be clipped over the pipework. Seal chinks in the insulation with a heavy-duty (preferably fabric-type duct tape) adhesive tape. Apply this along the split in the side, as well as to butt-joins between individual lengths. Extra tape will be needed around bends to hold the insulation in place and give a snug fit.

A cheaper alternative is glass or mineral fiber bandage. This is sold in rolls and you just wind it around the pipe. (Wear a filter-type face mask, gloves and a long sleeved garment for this – the fiber glass can irritate your skin and lungs.) Starting at one end of the pipe, tie the end of the bandage in place (use string or adhesive tape) then begin winding. As you work, secure the bandage at intervals with more string or tape and again take extra care when turning corners giving them an extra binding to be doubly sure. For the insulation to fit snugly without gaps, tie or tape it in position right around the bend.

Insulating tanks

Jackets for hot water cylinders are simply slipped around the tank and secured with straps. Since they are tailored to the job and supplied with fitting

Fitting a cylinder jacket; ensure that there are no gaps between the individual segments and that the insulation does not slip down inside the cover.

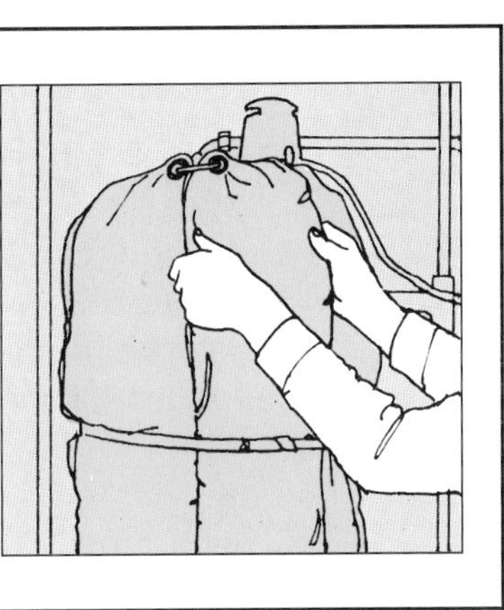

instructions, they should not present any problem at all. Measure the height and diameter of the cylinder before buying the jacket and ensure that it is manufactured to the standard set by your local building code.

Cold water tanks can also be fitted with pre-made jackets (often insulation blanket wrapped in tough plastic) but you can improve insulation for both the tank and its lid quite easily and cheaply. There are three main ways in which this can be done. The first is to swaddle the tank with the sort of glass fiber blanket used for attic insulation.

This is a messy job, however, and does not make it easy to get at the tank. A better method is to buy slabs of 1in thick expanded styrofoam. This can be cut with a sharp saw-edged knife, so it is a relatively simple task to fashion a casing for the tank, and to hold it in place at the joins using adhesive tape or cocktail sticks pushed through the corners of abutting panels.

If the tank is round, the simplest way out is to use a loose-fill material such as expanded styrofoam granules. Build an enclosure for the tank using hardboard or expanded styrofoam and pour the loose material in. For the lid, build a tray to take the insulant, or use an expanded styrofoam slab.

OVERHAULING THE CENTRAL HEATING SYSTEM

Before turning on your central heating at the start of the winter, it is always worth giving the system a quick overhaul. The boiler and pump should certainly be serviced professionally, and it is well worth taking out a service contract for the purpose. There are however, a few checks you can make yourself.

Start with the feed and expansion tank. Because it has very little to do, even during the heating season, the ball-valve here is far more likely to stick than the one on the main cold water storage tank, and this can lead to overflows as well as a failure to top up the system effectively. To prevent this, simply move the float arm up and down a few times to make sure it operates smoothly. At the same time, check that lowering the arm really does allow water into the tank, and that raising it shuts off the supply.

If the arm does have a tendency to stick, turn off the main stop-valve, and disengage the arm from the valve body by removing the retaining split pin. Clean off any corrosion at the valve end of the arm with wet and dry abrasive paper, grease the surfaces lightly with a little petroleum jelly, then refit the arm, and double check that it works.

Next turn your attention to the other valves in the system – both the on/off radiator valves, and any stop-cocks on the pipe runs. These, too, tend to stick if left unused for any length of time, so turn them all fully off, and then fully on to make sure they work. If they are stuck, try freeing them with a little oil.

Finally, double check the settings on any timers/programmers and thermostats. Turn the system on, and go round checking for leaks. The radiators are the most likely to fail. This is because the combination of steel radiators, copper pipes and water turns the system into a sort of electric battery, which is powered by literally dissolving the steel. Small leaks can be plugged with solder or with a suitable epoxy resin-based filler; badly affected radiators should be replaced, of course. To prevent further "electrolytic" corrosion, consider adding a corrosion inhibitor to the water in the feed and expansion tank.

Greasing the float arm of an expansion tank.

Adding corrosion inhibitor to the tank.

Bleeding a radiator valve

Finally, your radiators should be "bled" regularly, to release the air that builds up in the system. As this accumulates, the top half of an affected radiator tends to become cool. All you have to do is to insert a special key into the bleed valve at the top of the radiator, and turn it anti-clockwise, so opening the valve and allowing the air to escape. Close the valve, by turning the key clockwise, when the air has escaped and water appears.

The pump for the central heating system may also need bleeding if it seems sluggish in action. This is done in much the same way as for a radiator, except that the bleed valve can normally be operated with a screwdriver.

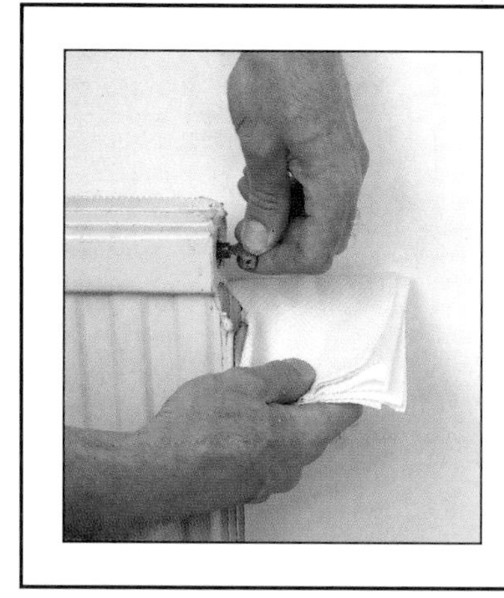

Bleeding a radiator. Turn off the pump and turn the radiator key a fraction of a turn counter-clockwise; close again when water appears.

Repairing burst pipes

A burst pipe requires quick and effective action to both minimize the damage done by escaping water and to get the system back in use as soon as possible.

Turn off the water supply to the affected pipe immediately and pack thick cloths around the damage to staunch the flow while you drain it down. If the pipe is damaged because you have driven a nail through it, leave the nail in place until the pipe is drained.

Once the pipe is drained, cut out the affected section and fit a new piece. You can buy a repair kit comprising a length of flexible copper pipe with two push-fit polybutylene connectors to allow a quick replacement to be made.

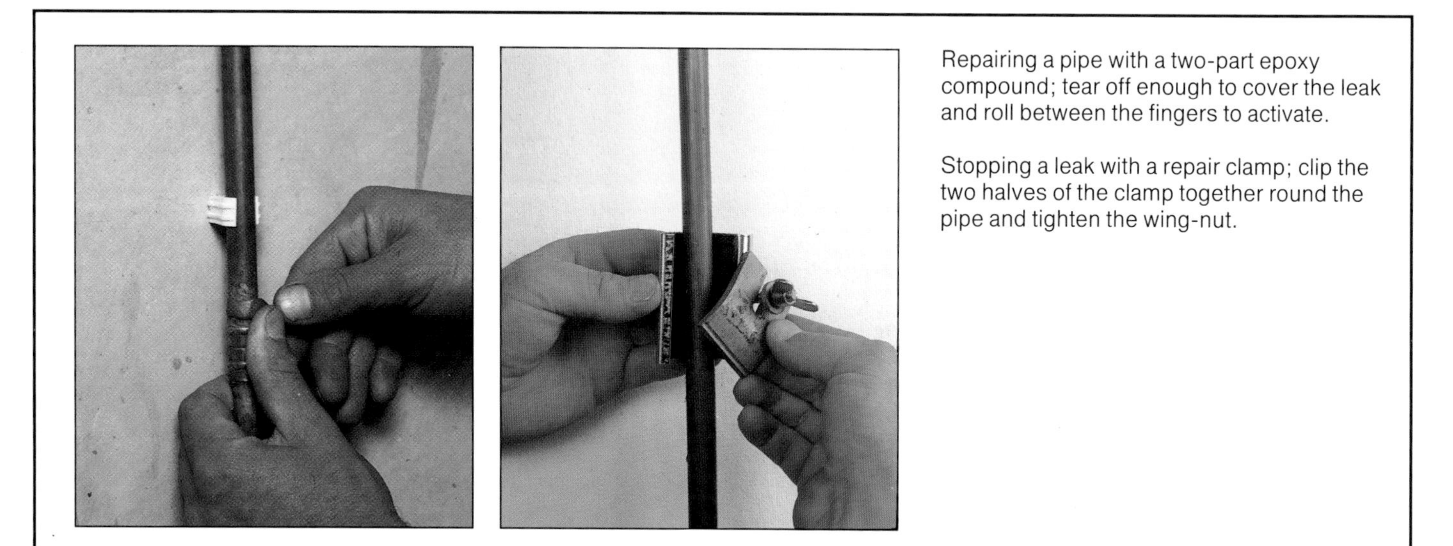

Repairing a pipe with a two-part epoxy compound; tear off enough to cover the leak and roll between the fingers to activate.

Stopping a leak with a repair clamp; clip the two halves of the clamp together round the pipe and tighten the wing-nut.

CHAPTER 2

INSTALLING THE SYSTEM

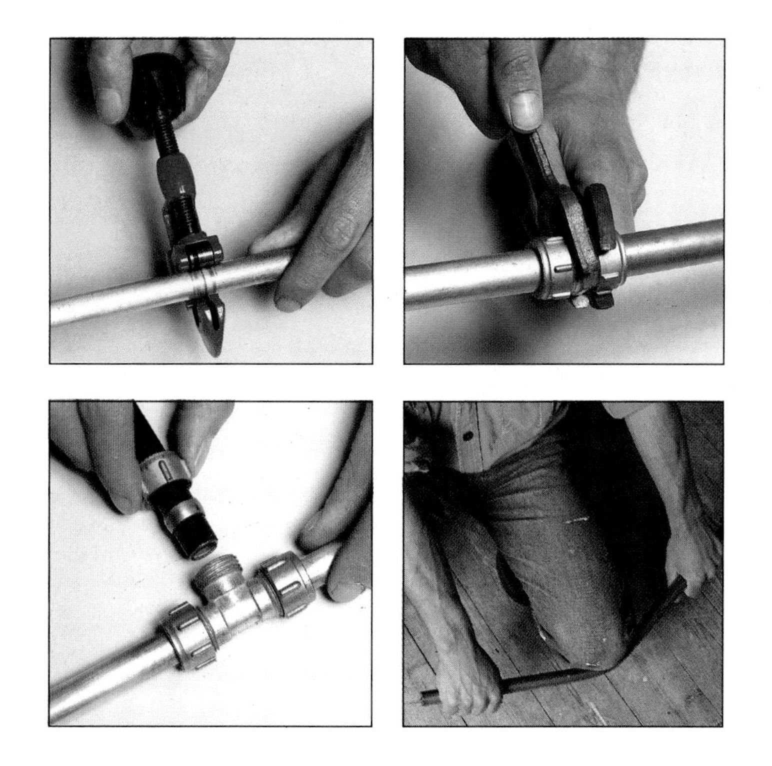

Using copper, or plastic piping in particular, it is now much easier to make repairs and extensions to the domestic plumbing system. Plastic piping is relatively cheap, light to handle, non-corrosive and does not adversely affect other materials. It can also be used for either hot or cold water supply, including central heating.

This chapter explains the techniques for handling both types of pipe; how to install a water filter; a new hot cylinder, and how to extend your central heating system. With plenty of professional tips, the home plumber is assured of achieving successful results.

• CHECKPOINT •

Using metal piping

The most common material used in modern plumbing systems is copper pipe which is easy to cut, bend and join. All copper pipe used for plumbing systems comes in imperial sizes, the most common being ½in, ¾in and 1in.

Measuring copper pipe

Lengths of copper pipe are joined by separate copper or brass connector or joint fittings where the pipe ends are inserted into them. Therefore, it is important when measuring up for a pipe length to make allowance for the margin of pipe inside the joint. Often the positions of the pipe stops inside the joint are marked on the outside but if they are not, it is a simple operation to find out how much pipe fits inside. Just insert a scrap of pipe into the joint and mark on it where the end of the pipe socket comes. Pull out the scrap and measure from the mark to the end. Incorporate this measurement in the overall length of pipe to be cut.

Cutting copper pipe

One of the most important skills to master when carrying out plumbing work is that of being able to cut a square end on a piece of pipe. A square end is essential if the pipe is to fit tightly against the pipe stop inside the joint, helping to form a watertight seal.

There is a simple way of marking a square cutting line around a pipe. First take a strip of thick paper with a straight edge and wrap it round the pipe so that the ends overlap and the edges are aligned. Then run a pencil around the edge of the paper; this will give you a square cutting line.

You can cut the pipe with a fine bladed hacksaw after first setting it in a vise. Do not overtighten the jaws of the vise as copper pipe is quite soft and easily distorted. Carefully saw through the pipe, keeping to the marked line. Then remove the burrs around the outer and inner edges with a half-round file. Make sure you shake out all the fillings as they could cause damage to faucets or ball valves if left inside. Devices are available for de-burring and brightening both the inside and outside of a cut pipe.

If you have a lot of pipe to cut, it may be worthwhile to rent or buy a wheel tube cutter. This ensures a square cut every time. To use it, all you have to do is to insert the pipe between its jaws and tighten up the adjuster. Then rotate the cutter around the pipe, tightening the adjuster as the cutting wheel bites into the pipe. The tool produces a burr-free outer edge to the

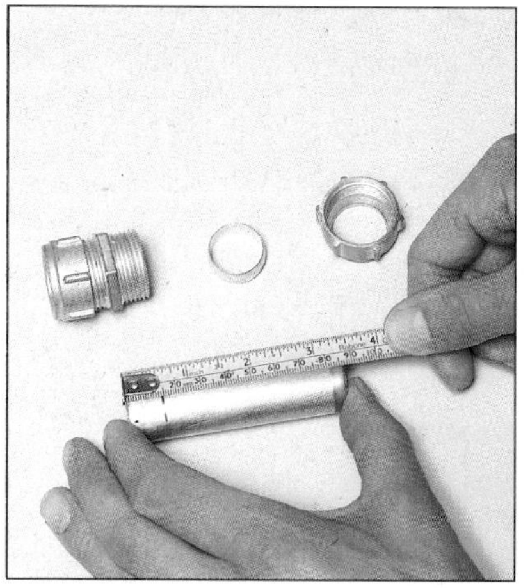

Measuring the depth of the pipe-stop in a fitting using an offcut; above, left to right: fitting, olive and capnut.

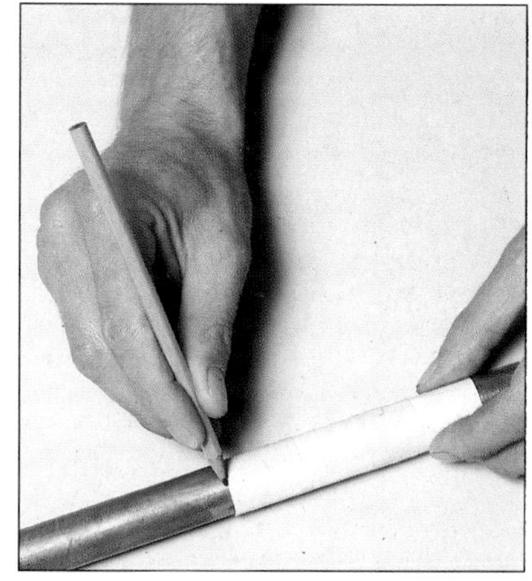

Marking a cutting line around the pipe against a piece of thick paper wrapped around it, to ensure a square end.

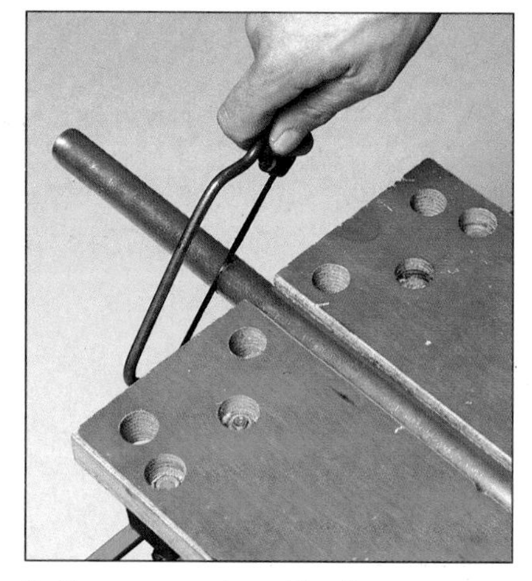

Cutting copper pipe with a fine-toothed hacksaw; a portable workbench is useful for holding the pipe without scoring it.

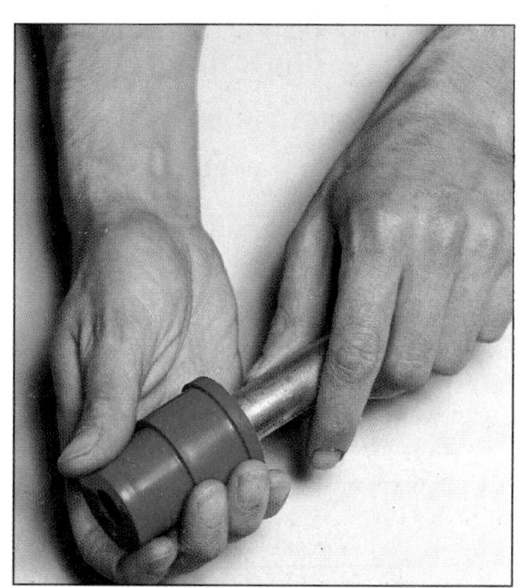

Using a de-burring brush to clean up simultaneously the inside and outside of a sawn end of pipe.

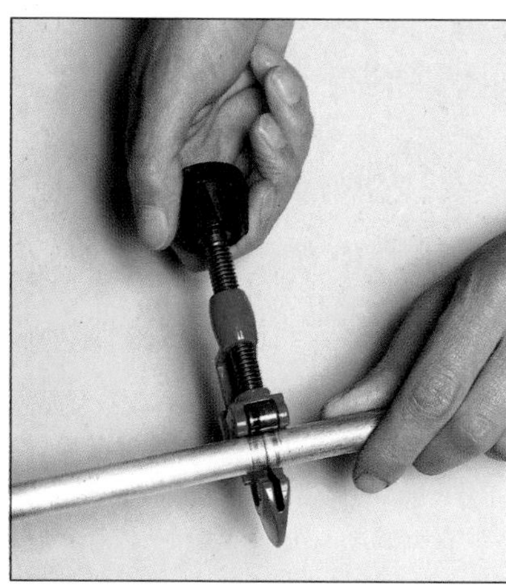

Cutting copper pipe with a wheel tube-cutter; rotate the cutter round the pipe, gradually tightening the screw-handle.

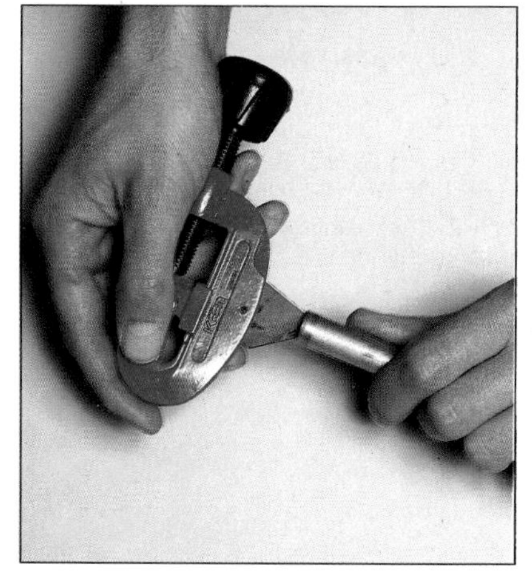

Reaming the cut pipe with the attachment on the cutter; wheel-cutters leave a burr only on the inside.

cut and usually incorporates a "reamer" (a hardened steel cutting edge) for scraping out the internal burr.

Bending copper pipe

Although you can buy angled joints for changing the direction of pipe runs, these are limited only to certain angles and they could be expensive if you had to use a lot in a pipe run. Where this is the case, it would be cheaper and neater to bend the pipe itself.

To do this, you will need a special tool to prevent the pipe kinking as you bend it. For ½in and ¾in pipes you can use a bending spring. This is pushed inside the pipe and positioned at the point where it is to be bent. Then you simply bend the pipe over your knee, the spring supporting the walls of the pipe. A length of stiff wire attached to the end of the spring allows its removal.

With 1in pipe, you will need a bending machine. The pipe is inserted between the formers on the machine and the handles are pulled together to form a bend.

Joining copper pipes

There are two types of joint used for joining copper pipes: the brass compression joint and the copper capillary joint. Compression joints are easy to fit and may be dismantled and remade at will, whereas capillary joints and soldered permanently in place; however, they are cheaper than compression joints.

The compression joint has a central body with a threaded collar or capnut at each pipe socket. A watertight connection is made by tightening the capnuts which, in turn, compress brass or copper rings (called "olives") between the body of the joint and the pipe.

A watertight seal is made with a capillary joint by melting solder between it and the pipe end. Some capillary joints have integral rings of solder and need only be heated with a blowtorch until the solder melts and flows round the pipe. Others, known as "endfeed" fittings, have to be heated then fed with solder separately. The latter are cheapest, but the former are easiest to use.

Bending copper pipe with an internal bending-spring; overbend slightly, bend back and pull out with the attached wire.

Using a pipe-bender; insert the pipe and the correct size of former and close the handles to give the required angle.

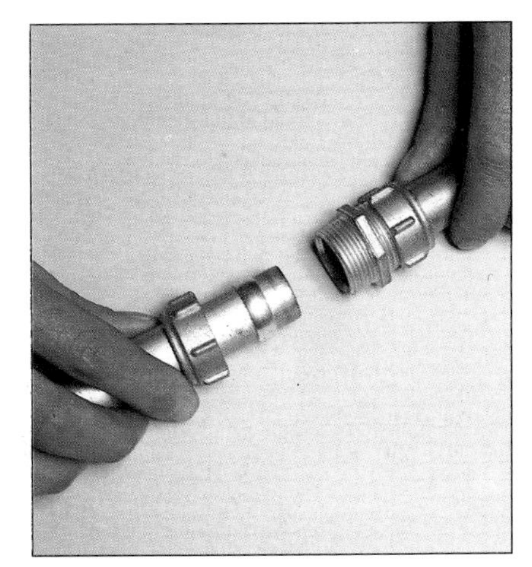

Joining copper pipe with a compression fitting; slip on first the capnut, then the olive, and screw into the joint body.

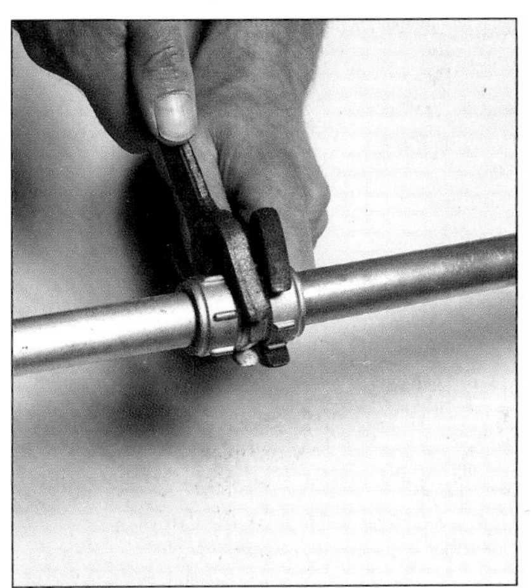

Tightening the capnuts with two wrenches; grip the joint body with one and give each capnut 1½ turns from finger-tight.

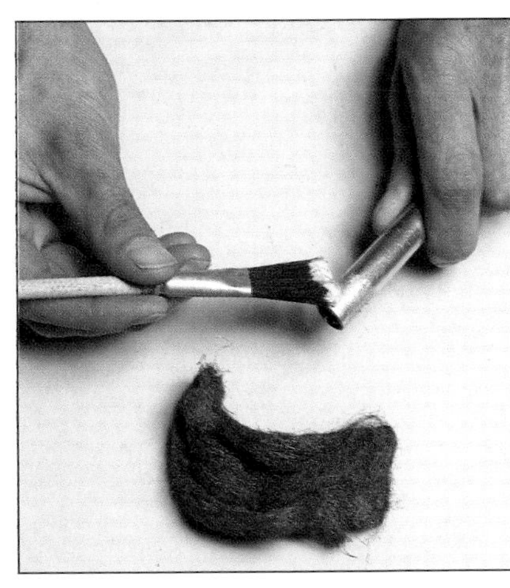

Applying soldering paste to the end of the pipe after cleaning with steel wool; use the correct soldering paste for the fitting.

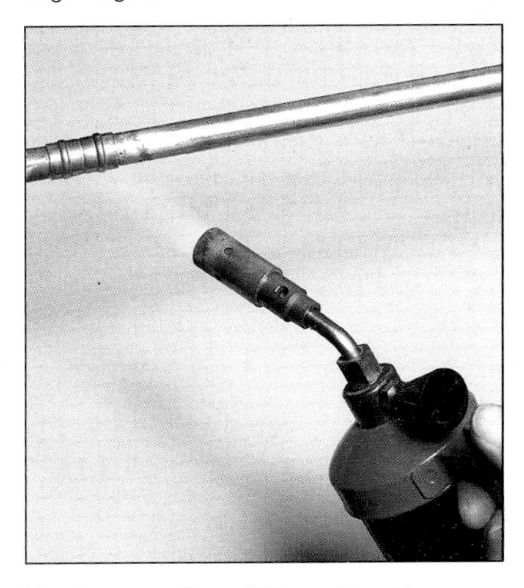

Heating a capillary fitting with a blow torch lamp; heat the whole fitting until rings of solder appear at the ends.

• CHECKPOINT •

Using plastic pipes

Plastic pipes have long been used for waste systems and for some time it has been possible to buy flexible polythene pipe for use with cold water supply systems; however, plastic pipe systems are also produced for both cold and hot water supplies, bringing the all-plastic plumbing system into reality.

Plastic pipework offers many advantages: it is light in weight, easily cut and joined, is a poor conductor of heat so less heat is lost from hot water and there is less likelihood of pipes freezing up. Also, it does not corrode, which can be a health hazard in some metal systems.

There are two types of plastic supply pipe available, but most common and easiest to work with is polybutylene pipe, a brown flexible material produced in ½in and ¾in sizes to match copper pipe. It can be used for any domestic hot or cold supply work with one exception – that is, it must not be connected directly to a boiler; you must fit short lengths of copper pipe first.

Polybutylene pipe

Polybutylene pipe is sold in 10ft lengths or 195ft coils and is flexible enough to be fed below floors or around obstructions. You can bend it by hand to a minimum radius of four times its diameter, provided you clip it on each side of the bend. It should also be clipped more frequently than copper as there may be a tendency for it to sag; 16in is the recommended spacing for the ½in size and 24in for the ¾in size.

Polybutylene pipe can be connected with normal brass compression fittings, making it easy to connect to existing copper systems. Or you can use the joints developed with the pipe. These look like plastic versions of compression joints, but there is no need to dismantle them to make the connection; you simply push the end of the pipe into the fitting until it comes up against the pipe stop. A toothed metal grab ring prevents the pipe from pulling out again and a rubber O-ring provides a watertight seal. Joints can be disconnected by unscrewing the capnut.

Unfortunately, this form of pipe is not approved by all Plumbing Codes (although it is increasingly finding favor), so you must check with your code before using it. Also, if your plumbing system was used to ground the electricity system (it is in many older houses) inserting plastic pipes will mean re-arranging the grounding; in this case, seek the advice of a qualified electrician.

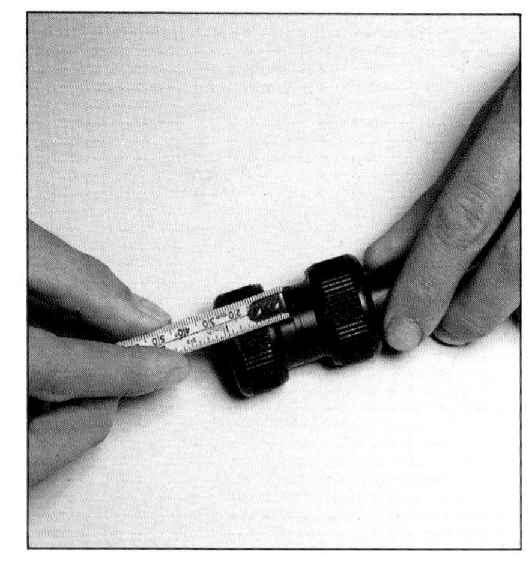

Measuring to the line marked on a plastic push-fit fitting; this is the distance to the pipe-stop within the fitting.

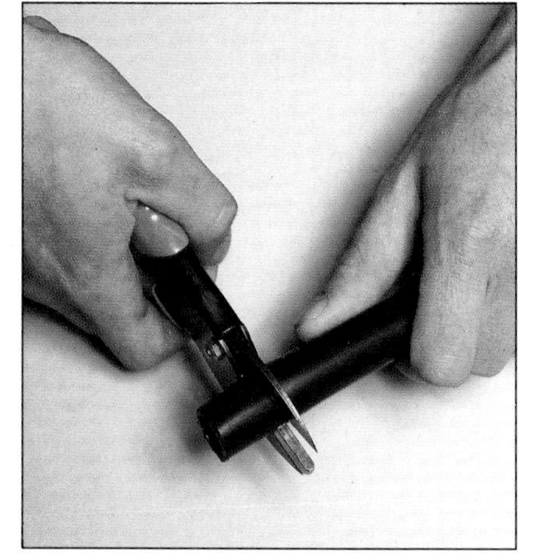

Cutting polybutylene pipe with secateurs; this method is quick and leaves no burr. A hacksaw may be used instead.

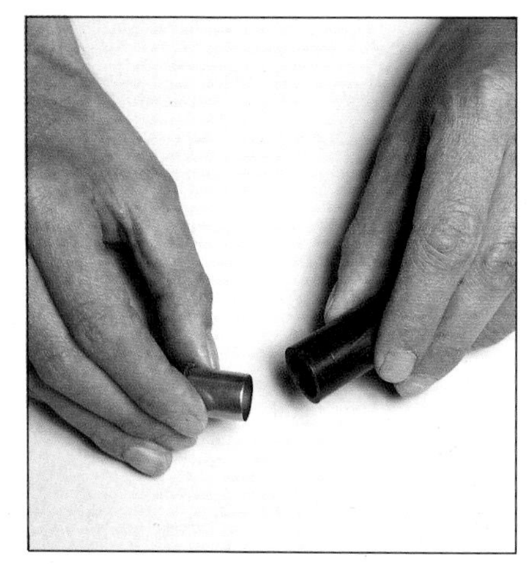

Pushing a stainless-steel sleeve into the end of the pipe to provide extra support where lengths are joined.

Smearing silicone lubricant around the end of the pipe to prevent chafing when it is pushed into the fitting.

Pushing the pipe into the joint; the knurled rings are pre-tightened and need no adjustment.

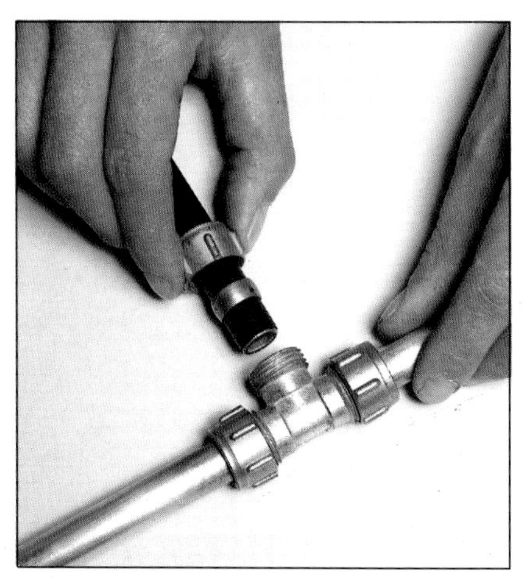

Joining to copper pipe with a standard compression fitting; slip on the capnut and olive before inserting the sleeve.

PLANNING PIPE RUNS

Although many appliances can be installed by connecting short lengths of pipe to adjacent supply pipes, sooner or later you will be faced with the job of installing longer pipe runs. A certain amount of pre-planning is needed if you are to avoid problems after installation.

Obviously, it is essential to keep the pipe runs as short and direct as possible, reducing the amount of work and time needed and also keeping costs down. It will help if you draw up a floor plan and sketch the pipe runs on it. Keep the changes in direction to a minimum together with the number of joints used. Joints are expensive and they are a potential source of leaks.

When you do have to change direction of the run, try to bend the pipe rather than using a right-angled joint, and where possible, keeping a gentle radius as this will not restrict the water flow as much.

Pipes can run horizontally or vertically, but avoid inverted loops of pipe as these are potential sources of air locks.

You must support pipes at regular intervals with proper pipe clips; this prevents any strain on the joints and stops water hammer caused by the pipes vibrating as water flows through them. Install the clips at 4ft intervals.

Running pipes under floors

In suspended wooden floors, pipes can be clipped to the sides of joists, or if the run is at right angles to the joists, the pipes can pass through notches cut in the joist tops or through holes drilled in them. Notches should be about ½in wider and ¼in deeper than the pipe, and holes ¼in larger in diameter to allow for expansion. Do not make them any larger in case you weaken the joist. Site the notches and holes clear of any floorboard fixing nails.

If you are installing the pipes below a wooden first floor, you may be able to clip them to the undersides of the joists, but make sure they are well insulated otherwise they may freeze up in winter (see page 18). Similarly, pipes can be run across the tops of joists in the attic, but here too they must be well-insulated.

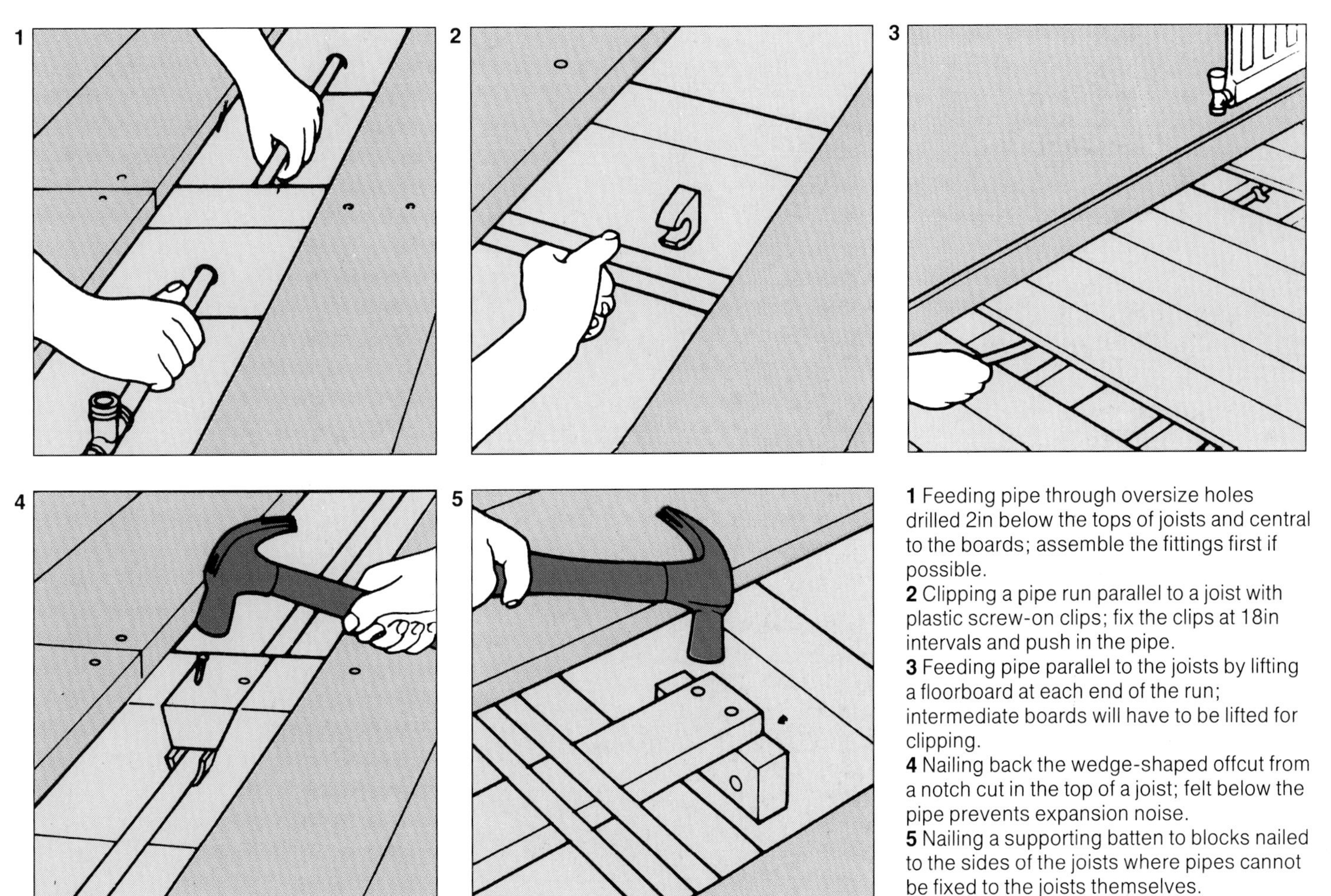

1 Feeding pipe through oversize holes drilled 2in below the tops of joists and central to the boards; assemble the fittings first if possible.
2 Clipping a pipe run parallel to a joist with plastic screw-on clips; fix the clips at 18in intervals and push in the pipe.
3 Feeding pipe parallel to the joists by lifting a floorboard at each end of the run; intermediate boards will have to be lifted for clipping.
4 Nailing back the wedge-shaped offcut from a notch cut in the top of a joist; felt below the pipe prevents expansion noise.
5 Nailing a supporting batten to blocks nailed to the sides of the joists where pipes cannot be fixed to the joists themselves.

Feeding pipe through access holes cut into a stud partition wall; drill oversize holes through bracing as squarely as possible.

Clipping a pipe in a chase cut in a solid wall; use saddle clips of the size to match the pipe and screw into wallplugs.

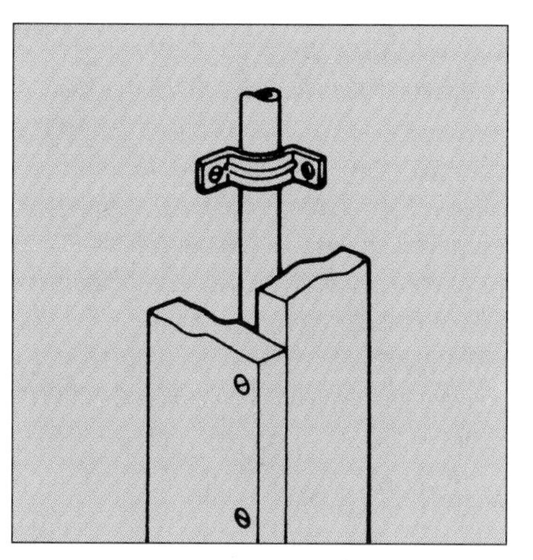

Surface mounted pipes, secured in a corner with angled saddle clips, can be boxed in with pine battens fixed through the edges.

Although you can bury pipes in concrete floors, installation is best done when the floor is laid or by casting a pipe channel in the floor which can be used later. Cutting a channel in an existing floor would be hard work and you will always run the risk of breaking through the waterproof membrane layer. It is probably better to clip the pipes around the walls at baseboard level.

Pipes in walls

Hollow stud partition walls are ideal for concealing pipe runs. If you are building the partition at the same time as installing the pipework, drill holes through the various frame members and feed the pipe through before completing the cladding (see page 26).

If the partition already exists, you can drill through the head plate from the room above and remove small squares of drywall in order to cut notches in the framework to accommodate the pipes. Then nail the squares back and fill the joints. You may be able to drill the sole plate from below after lifting a floorboard or two, or by running a long bit down behind the base.

In solid walls you can chisel a channel, or chase in the plaster deep enough to accept the pipe – special machines are available for rent to ease this chore – then clip it in place before plastering over the top. However, this is not recommended for hot pipes since their contraction and expansion may cause the plaster to crack. If it is necessary to run pipes along an exterior wall surface, then make sure that the wall is thoroughly insulated to prevent them from freezing. This way they will be less likely to freeze up in winter than if they were buried.

When running a pipe through one side of a wall and out the other, it is best to mortar a sleeve made from pipe of the next size up into the wall. This will then allow enough room for the pipe to expand without cracking the plaster. The sleeve should be cut just long enough so that its ends are flush with the plaster on each side.

INSTALLING A COLD WATER TANK

In those areas where, because of low rainfall or other reasons, it is necessary to store cold water on site, then you may find that they tend to be out of sight and out of mind; fortunately, they give very little trouble and, apart from the occasional check for corrosion in metal tanks and possible repair or replacement of a faulty filler valve, need little maintenance. However, there may come a time when you want to replace the tank you have; a common reason is that the old one does not have the capacity to meet your needs (most likely if you have installed extra appliances, particularly a shower). Alternatively, a galvanized steel tank may become badly corroded.

Choosing a new tank

Most modern tanks are made of plastic which is a much more suitable material than anything used previously. The ideal capacity for domestic use is 50 gallons and such a tank may be round or rectangular. However, before you buy, check the actual dimensions since the only way into the attic may be through the normal access trap and you must be sure you can get the tank through. Plastic tanks can be flexed considerably to pass through narrow openings, but if it looks as though there just is not enough room, you could buy two smaller units and connect them together in the attic with a length of 1in pipe. Fit the valve in one tank and take all the outlet pipes from the other to ensure a through flow.

Removing the old tank

Turn off the water supply and drain the tank by opening all the bathroom cold faucets. You will need a jug and bucket to scoop out the remaining water in the bottom of the tank.

Disconnect the pipework by unscrewing the connectors, but if they are difficult to remove, cut through the pipes and extend them later to meet the new tank.

Slide the old tank out of the way; if it is a metal one you may find it easier to leave it in the attic rather than attempt to manhandle it through the trap since it will be extremely heavy.

Installing the new tank

An important point to remember when carrying out any plumbing work in the attic is to always use compression joints or the newer push-fit polybutylene connectors, not capillary joints – the flame from a blowtorch could ignite dust in the attic with disastrous results.

A plastic tank, being quite flexible, must be stood on a firm base and this can be made by laying a few stout boards across the joists or using a sheet of ¾in thick plywood.

Holes will have to be drilled in the sides of the tank to take the connectors for the various outlet and inlet pipes, and their diameters should match the size of the connectors as closely as possible. The job is made easier by using a special hole saw or tank cutter fitting with an electric drill.

Fit the tank connectors, wrapping their threads with Teflon tape for a watertight seal. Then join the old pipes to them with short extension pieces. It is a good idea to take the opportunity of fitting gate-valves in the outlet pipes if none were fitted before, which will make repairs to the appliances they feed that much easier in future.

Fit the valve in the same way – either re-using the old one or a new one if it is unserviceable. Then add a plastic overflow pipe.

Turn the water back on and check for leaks as the tank fills. Finally, fit a lid, or make one from exterior grade plywood, and wrap the installation and pipes with a suitable insulation material (see pages 18–19).

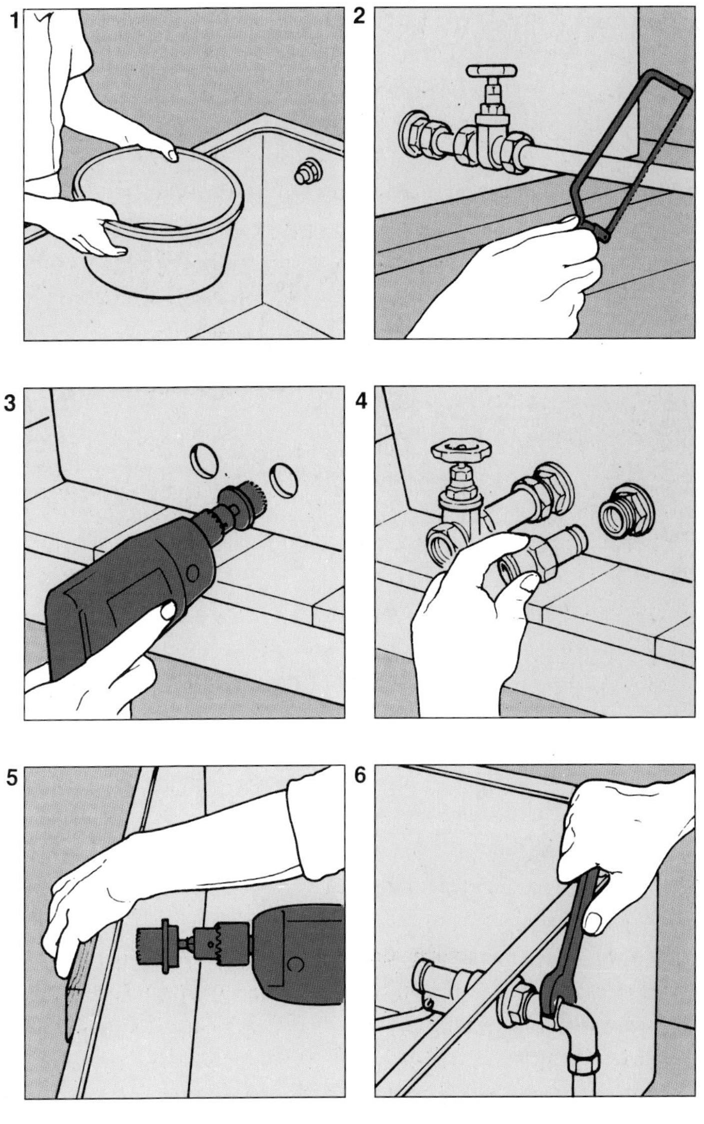

1 Removing the old cistern: baling out the water remaining below the feed pipes after draining the cistern by opening the bathroom cold faucets.
2 Cutting through the feed pipes with a hacksaw; carefully remove any burr so that it does not fall into the pipes and cause blockages to occur.
3 Installing the new cistern: cutting the outlet holes with a hole saw 2in above the bottom, to prevent residue of sludge from flowing in.
4 Fitting gate valves; a short length of ¾in pipe is needed to connect them to the tank connectors, using compression joints.
5 Drilling the hole for the supply pipe to the ball valve; support the wall of the cistern with a block of wood held behind the drilling point.
6 Connecting the rising main to the ball valve after wrapping PTFE tape round the ball-valve tail; clip the pipe firmly to prevent vibration.

WATER SOFTENERS AND FILTERS

In areas where the local water has a high concentration of mineral salts, you will notice the hard scale deposits build up inside water cisterns, pipes, washing machines and even coffee percolators. It is also responsible for discoloring ceramic baths and basins and can block shower heads. Most of these problems can be drastically reduced if not completely eliminated by installing a water softener.

The domestic water softener works on the principle of ion exchange whereby the incoming hard water passes through synthetic resin which absorbs scale-forming calcium and magnesium ions and, in exchange, releases sodium ions. After two or three days the softener automatically flushes the chamber with a saline solution to restore the resin. Granular salt is added to the softener every two or three months.

The units are electrically timed and are usually programed to operate in the early hours of the morning when there is the least demand made for household water.

Installation

The softener should be connected to the mains supply at the point where it comes into the house. The pipework should include valves and branch pipes both to supply and bypass the softener. The bypass will enable the unit to be isolated for servicing while the rest of the house is supplied with water.

You should also install a branch pipe to supply the kitchen sink with unsoftened drinking water, and possibly to a garden faucet. Your local authority will require a back flow preventer or check valve in the system, and perhaps a pressure reducing valve. You will also need a draincock to empty the rising main.

Many manufactures supply everything you need for installation in kit form and with full directions for assembling.

Water filters

A water filter is usually small enough to be attached to the pipe of a single main fixture, such as the kitchen cold water faucet. This would allow you to filter only the water used for cooking and drinking. Since the water filter must be installed in an upright position, where you actually put the filter affects the way it is installed.

Installing in a vertical pipe. If the water filter is to be installed in a vertical pipe (see the diagram above), cut out a 4 inch section of pipe, install a loop of piping with four elbows, and attach the filter in its lower leg. Place a gate valve on the inlet side of the filter. It is most important to install the filter in an upright position.

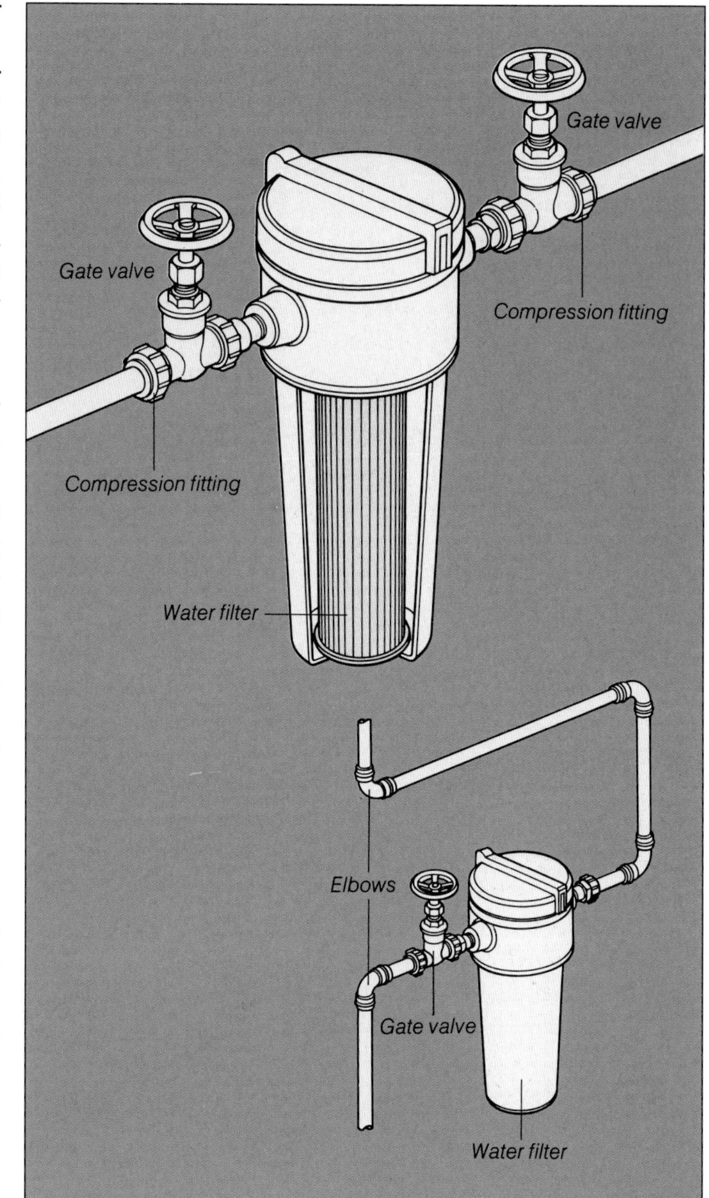

Installing in a horizontal pipe. To attach a water filter to a horizontal pipe (see the diagram above), cut out a length of pipe where the filter is to be installed. Thread a gate valve onto each side of the filter and attach the unit with compression fittings (or unions if galvanized pipe). Turn off both valves when replacing a used filter core.

To change a filter core, shut off the water by closing the gate valves. Unscrew the filter body from the cap and replace the used filter with a new one. Depending on the flow and the quality of the water, a filter core should last from six months to a year.

INSTALLING A NEW HOT CYLINDER

The supply of water to the hot faucets in your home will come from a copper storage cylinder. The water in the cylinder may be heated by an electric immersion heater, directly or by an oil or gas-fired boiler (see below).

There are two ways a boiler can heat the water in the cylinder: directly or indirectly. In the former, water is taken from the base of the cylinder, passed through the boiler and returned to the top of the cylinder where it is drawn off by the faucets. In the latter, there is a heat exchange inside the cylinder linked by a closed pipe circuit to the boiler. The water in this circuit is continually heated and circulated between boiler and heat exchanger which, in turn, heats the water in the cylinder. An indirect system is much better than a direct one since there is less chance of a build up of scale or corrosion in the boiler.

A hot water cylinder is unlikely to need much attention, but you may want to replace it with one of greater capacity or change a direct cylinder for an indirect one, or perhaps fit a cylinder with provision for installing an electric immersion heater. If your are installing one of larger capacity, make sure it has the same connections (direct or indirect) as the old one. The way to tell is that the boiler pipe tappings in a direct cylinder are almost always female whereas those in an indirect cylinder are usually male.

Removing the old cylinder

Turn off the immersion heater or boiler and allow the system to cool down before draining the cylinder, hot pipes and boiler circuit (see page 11).

Disconnect the pipework from the cylinder by unscrewing the various connectors and spring the pipes out of the way. If an immersion heater is fitted, isolate the circuit at the fuse panel. If you intend re-using the heater, you will need to buy or rent an immersion heater wrench to unscrew it.

Slide the old cylinder out of the way; being copper, it may be worth some money in scrap value which will help towards buying the new one.

Fitting the new cylinder

You may well find that the old pipes no longer match up to the connectors of the new cylinder, in which case you will have to cut them back and make up extension pipes, connecting them with capillary or compression joints.

Prepare the connectors for the cylinder by wrapping Teflon tape around their threads and screw them in tightly. The pipes are connected with compression fittings; fit the draw-off pipe to the top, the cold feed pipe to the bottom and the boiler pipes to the connectors in the side. Note that the feed pipe from the boiler is attached to the top connector and the return pipe to the boiler to the lower of the two.

If you are not fitting an immersion heater but the

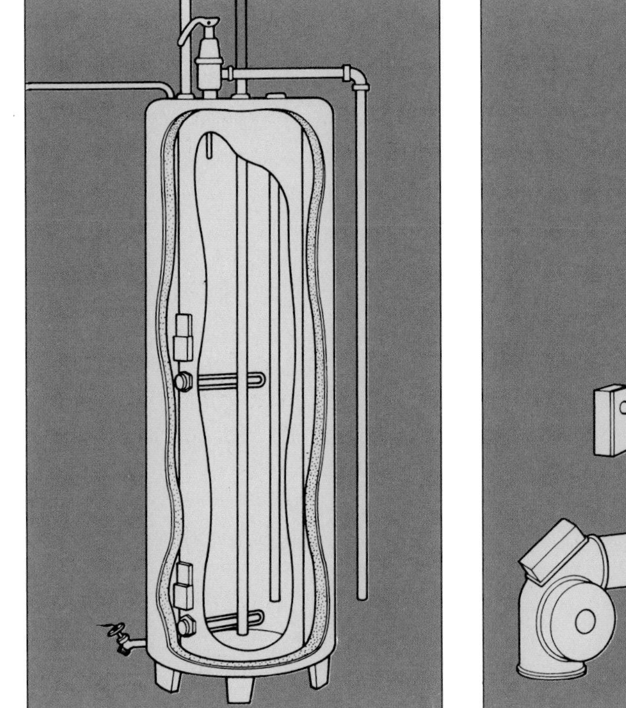

Electric hot water heater.

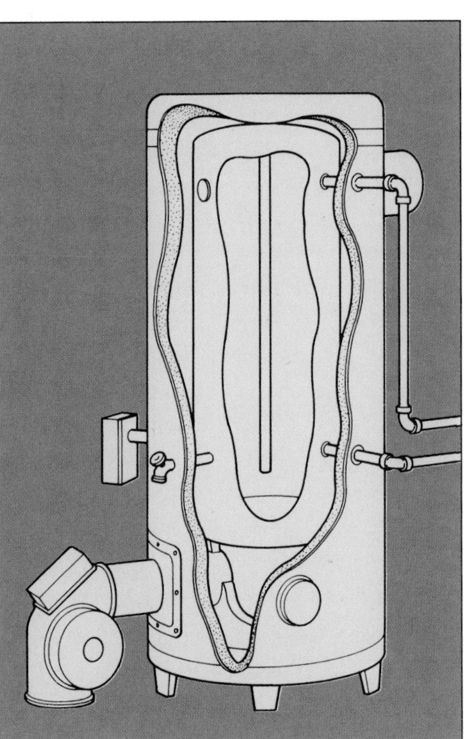

Oil fired hot water heater.

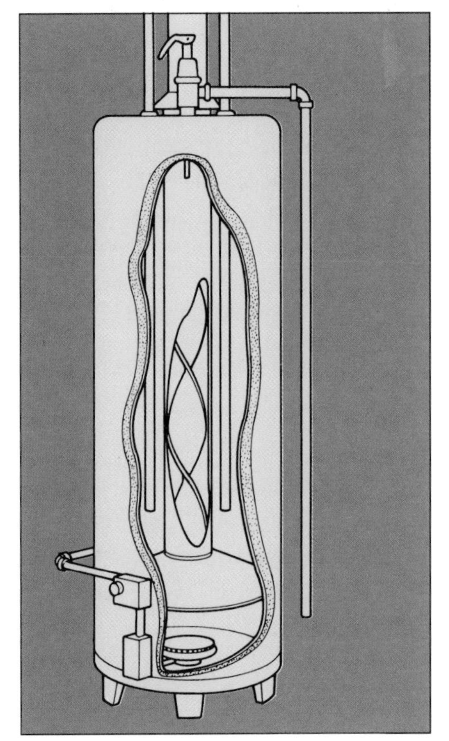

Gas fired hot water heater.

cylinder has a faucet for one, close it off with a special plate and seal the threads with tape. Similarly treat the threads of a heater unit, spread jointing compound over the faces of its sealing washer and screw it down tightly with the correct wrench. Then fit the thermostat and adjust its temperature setting.

Refilling the pipework

Refill the pipework and cylinder from the bottom upwards by connecting a hose between each drain-cock and the kitchen cold faucet in turn. This will prevent air locks. Check for any leaks.

Make the connection between the immersion heater and its fused circuit and turn it on – or start up the boiler – to heat the water. Check again for leaks; some may occur as the metal expands. If you find any, tighten the connections further.

Finally, if the cylinder does not already have molded-on or pre-fitted insulation, fit an insulating jacket.

EXTENDING YOUR HEATING SYSTEM

Extending your heating system is not in itself a difficult operation provided you have first checked with a reputable designer/installer that your boiler has the capacity to heat more radiators.

Draining down the system

Before doing repairs and alterations, you must drain the system of water, and it is worth knowing how to do this in any event, in case the system springs a leak.

The first job is to turn everything – especially the boiler – off and stop fresh water entering the system. Most systems are supplied from a small storage tank called a feed and expansion tank, filled from the main via a ball valve. This is usually situated in the basement near the boiler. To isolate it, look for a stop-valve on the main supply to the tank, or on the pipe leading from it, and turn it off. If there is not one, tie up the float arm of the ball-valve,or turn off the main stop-valve.

Now drain the pipework. There should be a drain-cock for the purpose. All you have to do is to connect a hosepipe to the spout on the drain cock, then turn the square boss on the top with a wrench. As water flows out, it can be directed, by the hose, into the garden, or into a sink. And where do you find the drain-cock? The short answer is: at the lowest point in the system – usually the boiler. Most freestanding models have a drain cock inside the casing at the front. If not, try along the pipes between boiler and radiators, bearing in mind that a system with two or more distinct radiator circuits, may have a drain-cock for each.

Can the system cope?

One of the most important considerations is to establish whether your particular system can cope with extra radiators, and this involves some rather complicated calculations. Even if the pump can get water to the new radiator – which may not be possible if this is in an attic – the boiler may not be up to heating it properly.

One way out is to contact a firm specializing in "central heating parts". Although these firms are usually geared to supplying complete systems, they may be willing to plan extensions, and supply the materials you need. You will find that many systems are installed with some spare capacity and you may be tempted to add only one or two radiators, but this would be unwise without first checking with a heating specialist. Should your boiler not have the extra capacity needed, any extra demands made on it would not only overtax the thermostat but would further reduce the general heat output – and thus defeat the purpose.

In this case, your heating specialist would probably advise installing a bigger boiler.

New pipes for new radiators

If the existing system is run in standard copper pipe, installing new radiators should be fairly easy. Begin by tracing the feed and return circuit pipes supplying existing radiators, and decide where to tap into them. If you are adding only one radiator, any convenient point will do. Break into each pipe with a compression or capilliary tee fitting, and run branch pipes to the new radiator connection.

To create what amounts to a new radiator circuit, tap in at the end of an existing circuit, or close to the pump on the side farthest from the boiler. Using tees to break into the feed and return pipes, run a pipe from each one along the shortest practical route between new radiators, teeing branch pipes off for connection to the new radiators.

Connections

Radiators come with a hole at each corner. The bottom two take an on/off valve and a balance valve, controlling the rate at which water leaves the radiator. Of the top two, one is merely plugged, while the other houses the bleed vent.

Both valves are fitted to the radiator using tail pieces (often supplied with the valve). Wrap their threads in Teflon tape to ensure a watertight join. Tape is also needed around the thread of the radiator plug, but here you may have trouble tightening up, since the plug will normally have only a square or octagonal recess by which it can be gripped. A "lever bar" is needed here. However, before going to the expense of buying one, try using a large screwdriver.

Finally, fit the bleed vent. How you do this depends on its design. If the part turned by the vent key is encased in an outer sleeve, bind the thread on the sleeve with Teflon tape, then screw it into place with a wrench. If, you have a one-piece vent, the whole of which screws in and out of the radiator, use the vent key to drive it home. Do not use tape or jointing compound on its thread.

You can now hang the radiator. It hooks onto special brackets screwed to the wall, but first make sure the wall fixings are firm – a full radiator is extremely heavy. In solid masonry, ensure that wall plugs expand in the body of the wall; not in the plaster. On stud partition and lath-and-plaster clad walls, do not use cavity fixings. Screw into the wooden studs or into battens screwed to bridge two or more studs.

Care is also needed when positioning the brackets, if the radiator is to be in the right place. Measure the distance between the centers of the fixing lugs on the back of the radiator, and the distance between the lugs and the top and bottom edges of the radiator. From these you can work out where the brackets should go for any given radiator position, and so long as you check that the brackets are vertical all should be well.

Finally, connect the feed and return branch pipes to the radiator valves. This is normally done using compression joints (see pages 23–24). Connect the feed pipe to the on/off valve and the balance valve to the return.

Moving a radiator

Repositioning an existing radiator is another job you can easily do yourself.

If the old and new positions of the radiator are fairly close, just extend the existing radiator branches of the feed and return circuits to the new position using standard plumbing fittings. Where the two positions are widely separated, cut the branch pipes back as far as they will go, then seal them off with stop-end fittings. The radiator can then be fitted in its new site, in the same way as if it were new.

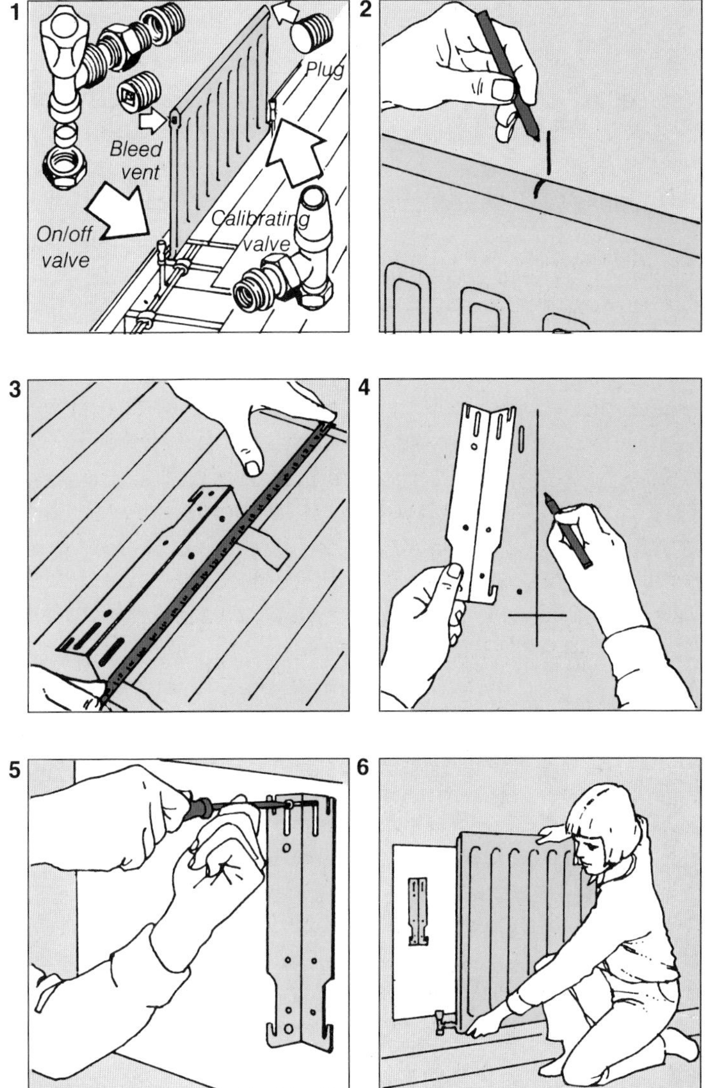

1 Components needed for a radiator: bleed vent; blanking plug; on/off (or thermostatic) valve; balancing valve.
2 After marking the center line of each hanger on top of the radiator, transferring the marks to the wall as a guide to fixing.
3 After positioning one bracket on a hanger, measuring from the top of it to the bottom of the radiator; add to this the floor clearance.
4 Marking through the fixing holes onto the wall with the bracket positioned to the calculated height above the floor.
5 Screwing the bracket to the wall, using 2in screws and wall plugs, with the corner of the bracket against the center line.
6 Lifting the radiator onto the two level brackets; plastic sleeves pushed onto the bracket lugs will prevent expansion noise.

Recommissioning the system

After such major changes to the central heating, it is necessary to recommission the system, and the first job is to refill it with water by reversing what you did to isolate the feed-and-expansion tank during the draining process. Leave the drain-cock open with hose attached for several minutes so that the fresh water will flush out any debris, then close it and let the system fill. Once full, bleed the pump plus all radiators to release any trapped air.

If you have added a new radiator, the system will now need rebalancing. This may also be necessary if you have moved a radiator some distance. It is basically a matter of ensuring all radiators give out just the right amount of heat, and this is done by adjusting the balance valves using a special key (you can buy one cheaply) the more open the valve, the faster water will pass through the radiator, and the less heat it will lose.

You will have to rent special thermometers which can be clamped to the pipes for this. To balance the entire system, fully open all balance and on/off valves, turn on the boiler and allow about half an hour for the system to reach temperature. With the aid of the thermometer, note the temperature of the return pipe on each radiator and compare it with the temperature indicated by the boiler thermostat.

One radiator – the index – will have a far lower reading than the others, and this should be left alone. For the rest, work round the system gradually closing down the balance valves, and renoting the temperatures, until all are within a degree or so of each other. Finally, check the temperature of the index radiator, if this is wildly different from the rest, try adjusting the water pressure ("head"), provided by the pump, assuming this is of the variable head type.

If you think all that sounds like a lot of hard work, just for the sake of a fairly minor alteration to the system, you are right. What is more, it requires a fair amount of experience to balance a system properly, within a reasonable time – it could take you days. So, is there a short cut? Why not just balance the radiators you have added or moved?

Well, if you have simply moved a radiator, or added only one or two, it is certainly worth a try. By closing down the balance valves on the new additions and opening the valves on any existing radiators farther along the feed and return circuits, there is no reason why you should not achieve acceptable results. However, as a final test, do carry out a temperature check on all radiators in the system, just to make sure everything works.

Thermostatic radiator valves

Instead of the ordinary on/off valves, thermostatic valves can be fitted to control the flow of hot water to each radiator, and thus control the temperature of individual rooms – which can sometimes be a hit-and-miss affair, since a room thermostat in, say, the living room, can hardly know whether a bedroom is too hot or too cold. They make alterations to the heating system simpler as you would no longer have to rely on complicated permutations like the size and number of radiators in a room to regulate its temperature.

If they have a drawback it is their cost. Fitting them to every single radiator in the house can be rather expensive.

• CHECKPOINT •

Fitting a thermostatic radiator valve

Drain down the system (see page 13) and remove the old on/off valve plus the radiator tail pipe and coupling nut. Fit the tail pipe supplied with the new valve, binding its thread with Teflon sealing tape to ensure a watertight join, then connect the thermostatic radiator valve to this tail pipe, tightening up the coupling nut. Finally, connect the radiator supply pipe to the new valve, shortening it if necessary, and fitting it with a new olive (see page 23). Recommission the system, and set the new valve for the required room temperature.

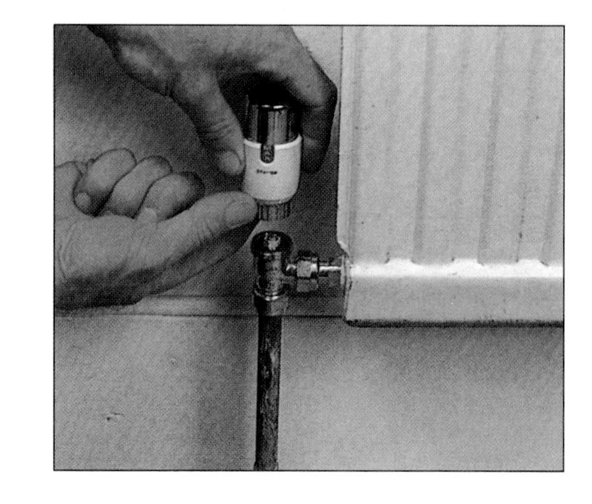

CHAPTER 3

WORKING WITH THE WASTE SYSTEM

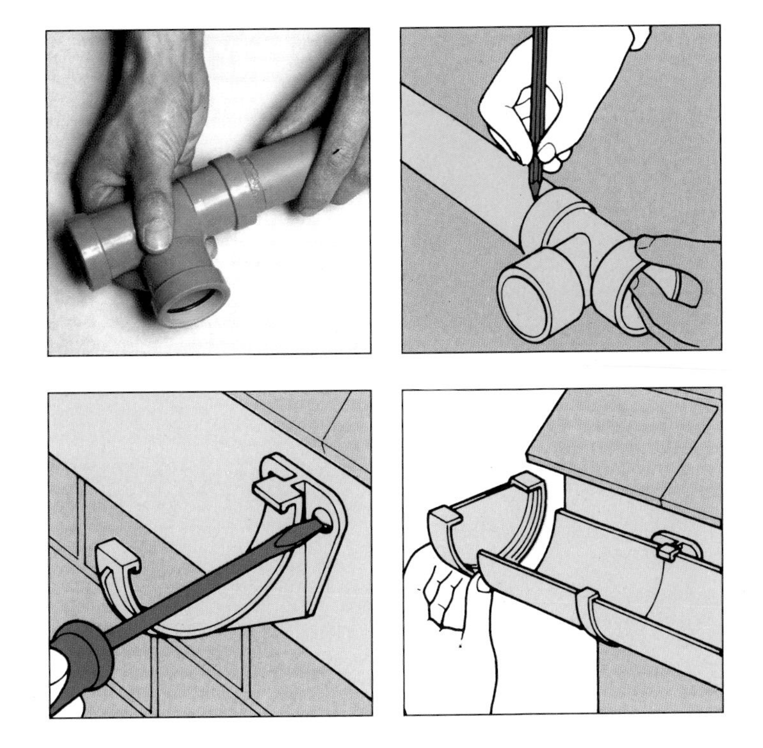

To ensure that the plumbing system in your home runs smoothly at all times, it is important to know that the waste pipes can cope with your existing arrangement, and what effect any new appliances would have on it.
This chapter explains the intricacies of waste pipe runs — the drainage system designed to carry all domestic waste from the house through underground drains to the main sewer; how to install waste pipes and make watertight joints. Information is also given for maintaining your gutter system (which feeds into a separate drain), and how to replace it with a plastic system, for greater efficiency and protection for your home.

PLANNING WASTE RUNS

Installing new appliances should not cause any major problems if you are replacing existing fittings and simply reconnecting to the original supply and waste pipes. However, if you intend changing the position of an appliance or installing an extra one, the problems begin. The supply side of the job is quite straightforward, but the way you tackle the waste system is more complicated. Actually installing the pipework is not difficult, but waste systems must comply with the Building Code.

Before you tackle the job you must draw up a set of plans to show the pipe runs and submit them to the local Building Department for their approval. Following this they will probably want to see the installation when it is complete.

Single stack and two pipe drainage

Your house may have a modern single stack drainage system in which a large diameter pipe, connected to the underground drain, runs up an outside wall or, more likely in modern houses, through the house inside a duct. All the waste and soil pipes from the upper floor appliances and fixtures will be connected to the stack, and some ground floor appliances may also be linked to it. However, it is quite possible that a downstairs toilet will be connected directly to the underground drain.

Running waste pipes

Waste pipes are larger in diameter than supply pipes, particularly soil pipes which measure 4in, and nowadays are nearly always made of either plastic of hubless cast iron with slide fitting connecting clamps.

Because of their size, waste pipes are much more difficult to conceal than supply pipes. Keep the pipe runs short by siting appliances as close as you can to the stack position. You can, of course, box in waste pipes and even run them between the joists of a wooden floor. However, you must never cut the joists to allow a pipe (even the smallest diameter waste pipe) to pass across them as this will seriously weaken the floor. In the situation you have no alternative but to run the pipe along the wall, clipping it at regular intervals just as you would a supply pipe.

Waste pipes must also have a positive fall towards their discharge point – about 1¼in per yard.

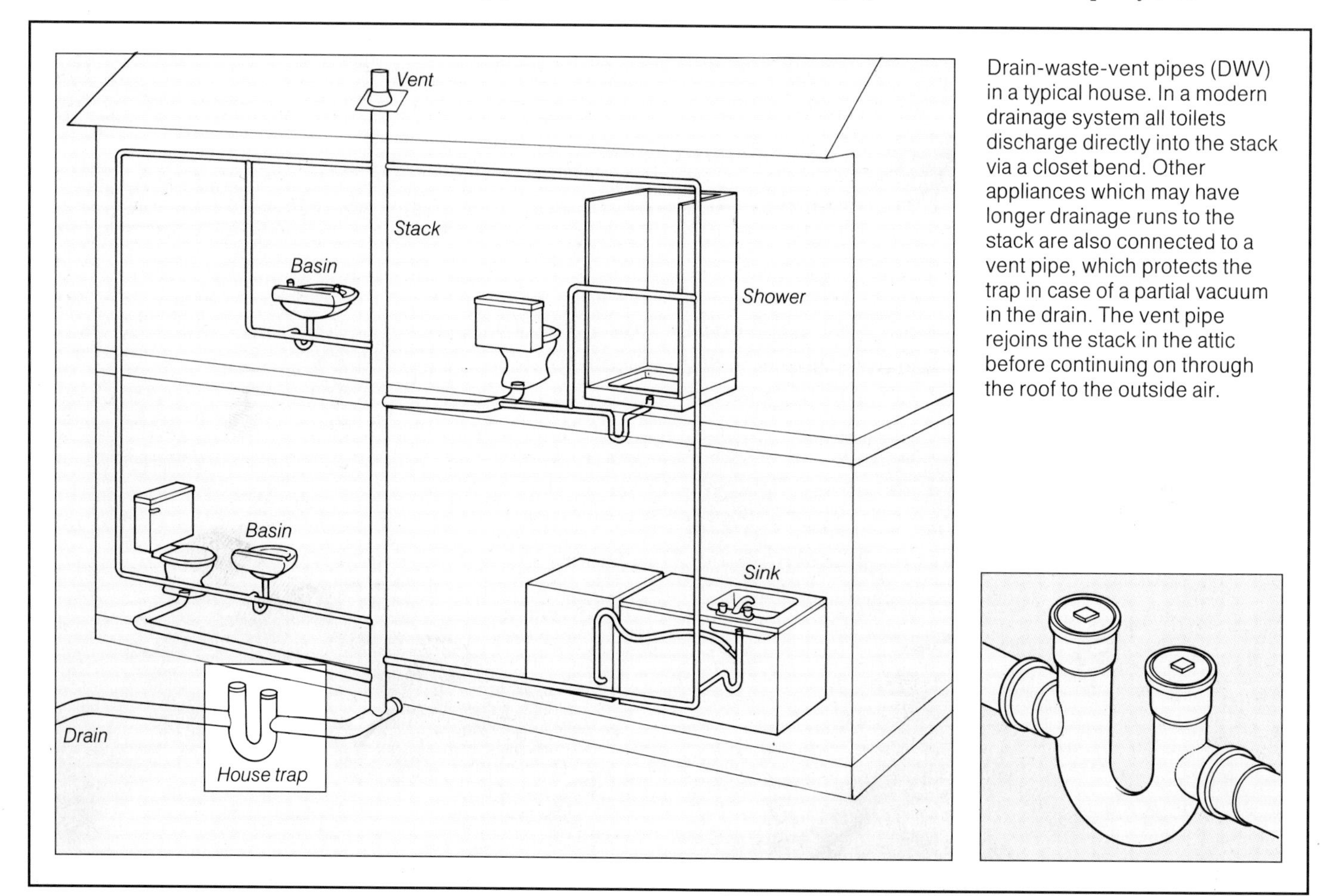

Drain-waste-vent pipes (DWV) in a typical house. In a modern drainage system all toilets discharge directly into the stack via a closet bend. Other appliances which may have longer drainage runs to the stack are also connected to a vent pipe, which protects the trap in case of a partial vacuum in the drain. The vent pipe rejoins the stack in the attic before continuing on through the roof to the outside air.

Choosing the pipe size

You will find several sizes of waste pipe in use: 1¼in, 1½in, 2in, 3in, 4in are common, and possibly 6in. The larger sizes are for use with toilets and for vertical waste stacks only. The only appliances that use 1¼in pipes are a hand basin and a bidet and then only if the pipe run is less than 5ft 7in long; above that figure, up to a maximum of 7ft 6in, use 1½in pipe. The waste pipes from baths, showers and sinks must all be in 1½in pipe.

Making the connections

Where possible, avoid making direct connections to the soil stack; rather, connect new appliances to existing waste pipe runs using T connectors.

However, the positions of existing appliances may not always make connection to their waste pipes possible, in which case you have no option but to connect to the soil stack, provided it is plastic. Basin and bath wastes can be solvent-welded to a spare entry boss on the stack after cutting out the circle of plastic inside, or joined to it with the aid of a "strap boss" fitting.

If you have to connect an extra toilet to a soil stack you may need to replace the existing single branch fitting with a double branch – complicated, since it means cutting out the old one and making up the missing pipe sections before solvent-welding the new one in place.

If the stack is of cast iron a section can be removed and replaced with a hubless cast iron T fitting, using the slide fitting connecting clamps. Merely position the new fitting, mark the pipe and cut it out, slide the connectors over and tighten. Two other points to consider are that you must not connect any waste pipe any closer than 8in below a soil pipe that joins the stack on the opposite side; and all connections to the stack must be at least 18in above the point where it bends to join the underground drain.

For waste disposal units, which are usually designed to fit a 3½in outlet in the base of a sink bowl, the discharged waste slurry is washed either into a (ground level) yard gully or soil stack, see pages 11 and 34.

WASTE PIPES AND FITTINGS

Waste pipes come with internal diameters of 1¼in and 1½in bores respectively. The smaller size is only used for short runs from hand basins and bidets otherwise all waste systems are in the larger size. Soil pipes come in 3in and 4in sizes, although you may find that older cast iron pipes are slightly larger; fortunately, you can buy special push-on adaptors.

It is not a wise idea to try to bend plastic waste or soil pipes but since there is a wide range of straight connectors, bends or various angles and T-junctions this is not a drawback. It is worth having a good look at the range of systems available before you start to see which have the fittings that will suit you best.

There are two methods of joining plastic waste pipes: solvent-welding and push-fit connectors. The former makes a permanent seal between pipe and connector by means of a special solvent, which is brushed on to both pipe and connector, melting and fusing the plastic together. The latter has simple rubber O-rings to make a watertight seal and can be dismantled and remade if required.

An advantage of the push-fit system is that an allowance can be made at each joint for thermal expansion caused when hot wastes flow down the pipes; this cannot be done with solvent-welding, so on long runs special expansion joints must be incorporated.

Push-fit connections

Push-fit joints are easily made. Bevel the pipe end with a file and apply a coating of petroleum jelly to lubricate it. Push it into the joint socket and then pull it out again by about ¾in to allow for thermal expansion of the pipes.

Many old houses may still have copper or lead waste pipes and traps running from their baths and basins, together with cast iron soil pipes from the toilets. However, today, plastic is the universally accepted material for both waste and soil pipes, and this is much easier to work with than its metal counterparts.

Various types of plastic are used for waste and soil systems: polyvinylchloride (PVC), acrylonitrile-butadiene-styrene (ABS) and polybutylene (PB). All

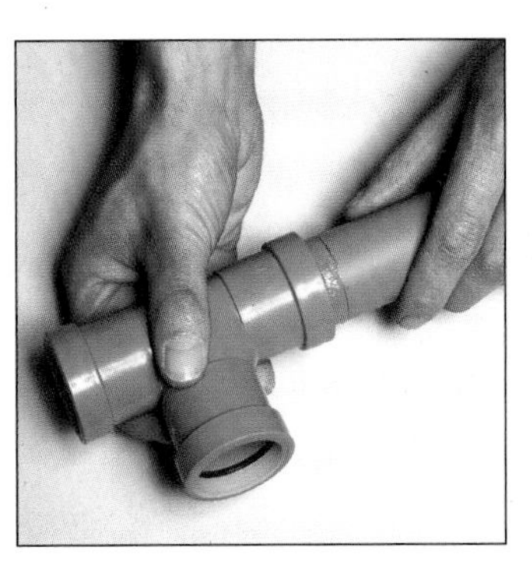

Making a push-fit joint; push in pipe fully, mark round mouth of socket, then withdraw pipe ⅜in to allow for expansion.

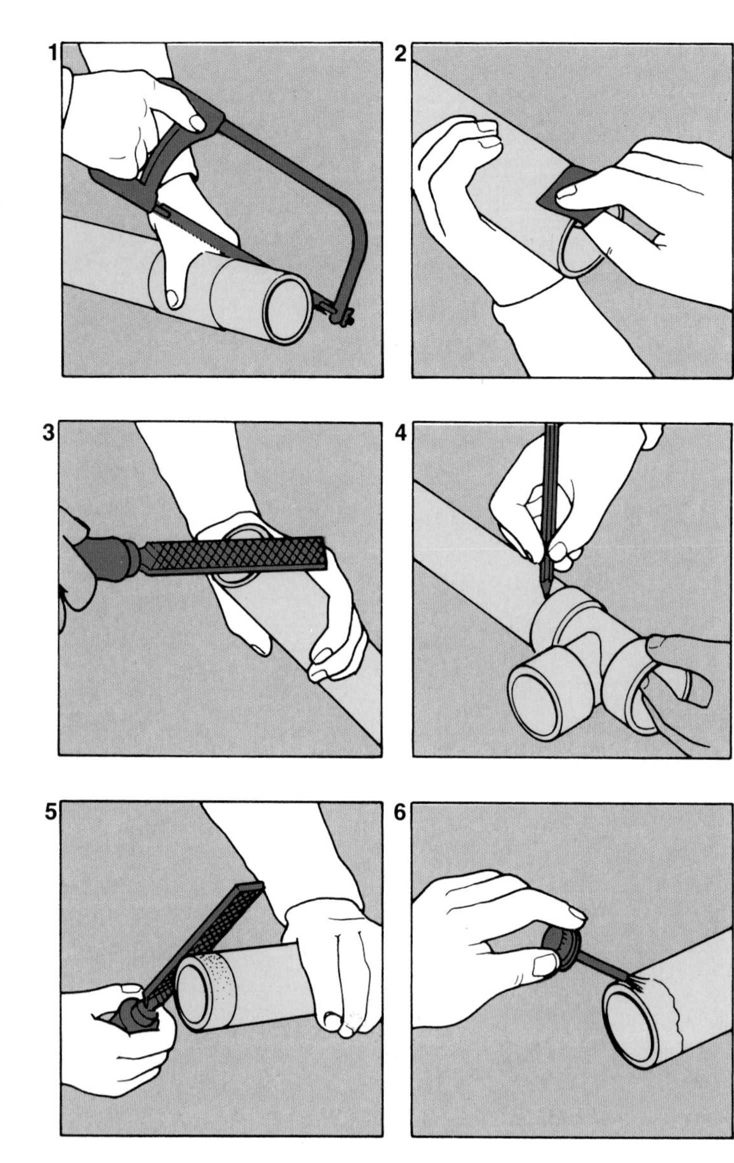

1 Making a solvent-weld joint in waste pipe: cutting the pipe to length using the edge of a piece of paper as a cutting guide.
2 Removing burrs from the cut end with fine abrasive paper held lightly; alternatively use a fine metalworking file.
3 Chamferring the cut end with a file to give it an easy fit in the pipe fitting; make sure that you do not file the end out of square.
4 After wiping with a rag to remove all loose particles, and inserting fully into the fitting, marking around the socket as a guide.
5 Roughening the end of the pipe within the marked line; use a fine file or wire wool, and roughen inside the socket also.
6 Applying an even layer of cement to the pipe after cleaning; apply to the fitting also and push in the pipe with a twisting motion.

are suitable for domestic use so it does not matter which you choose with the proviso that you use the same material and brand throughout the system. Different types of plastic will not be compatible with each other, and different makes may vary slightly in size so that they cannot be connected together to guarantee a watertight seal.

Making solvent-weld joints

As with connecting copper or plastic supply pipes, it is essential to cut the ends of pipes squarely so that they butt up evenly against the pipe stops inside the connectors. Measure each length of pipe, making an allowance for the amount inserted into the connector, and use the same method to mark a square cutting line as that described for copper pipe on page 22; that is, use a straightedged strip of paper wrapped around the pipe with the ends overlapping and the edges aligned. Butt a pencil up to the paper and run it around the pipe.

Having marked the line, you can cut the pipe with a fine-bladed hacksaw. File off any burrs and, to aid insertion, use the file to bevel the edge of the pipe. Push the end into the joint socket as far as it will go and mark the pipe with a pencil so you will know how much of the end to prepare.

Remove the pipe and use a file or steel wool to roughen the end of the pipe up to the pencil mark. Similarly, treat the inside of the joint socket with steel wool until you have removed the "glaze" from the surface of the plastic. Then wipe the end of the pipe and inside the joint socket with the appropriate pipe cleaning fluid (as recommended by the manufacturer).

Make sure you have the correct solvent-weld cement for the type of pipes you have; different plastics use different cements. Then, using the brush supplied with it, or a small clean paintbrush, apply the cement to the end of the pipe and the inside of the joint socket.

Push the pipe into the socket until it comes up against the pipe stop, twisting it slightly as you do to spread the adhesive. Then hold the pipe and joint together for 20 to 30 seconds to give the adhesive a chance to make the bond. Wipe off any excess, and make sure that the connector points in the right direction for the next length of pipe. If the pipe is to carry hot waste water, leave it to set for 24 hours before using the system.

YOUR GUTTER SYSTEM

Your roof presents a very large flat surface to the sky and during a rainstorm the amount of water flowing off it can be quite considerable. If nothing was done to collect this water, it would simply cascade down the walls of your house, seeping into the brickwork and doing untold damage. It would not be very pleasant for anyone walking underneath either. That is why every house has a rainwater collection system.

The rainwater system comprises gutters which run along the roof at eaves level to catch the water as it runs off and channel it to vertical downpipes. The downpipes either discharge the water into a trapped gully at ground level or are connected directly to an underground drainpipe. This, in turn, may direct the water to a storm drain running under the road outside, or to a soakaway in the garden.

Rainwater system materials

Early rainwater systems were made of copper or wood and many houses still have such systems. However, wood is not the best of materials to use for something that is in constant contact with water; it can suffer from rot if not painted regularly.

These days aluminum and plastic are the most common materials for rainwater pipes and gutters: they are light in weight, unaffected by corrosion, do not adversley affect other materials, require virtually no maintenance and are easily installed.

Both systems offer a choice of three basic gutter profiles: half-round, semi-elliptic and square, and common sizes are 3in, 4in and 6in, these being measured across the widest part of the gutter. They are available with matching downpipes and a wide range of accessories that allow even the most complex arrangements to be duplicated.

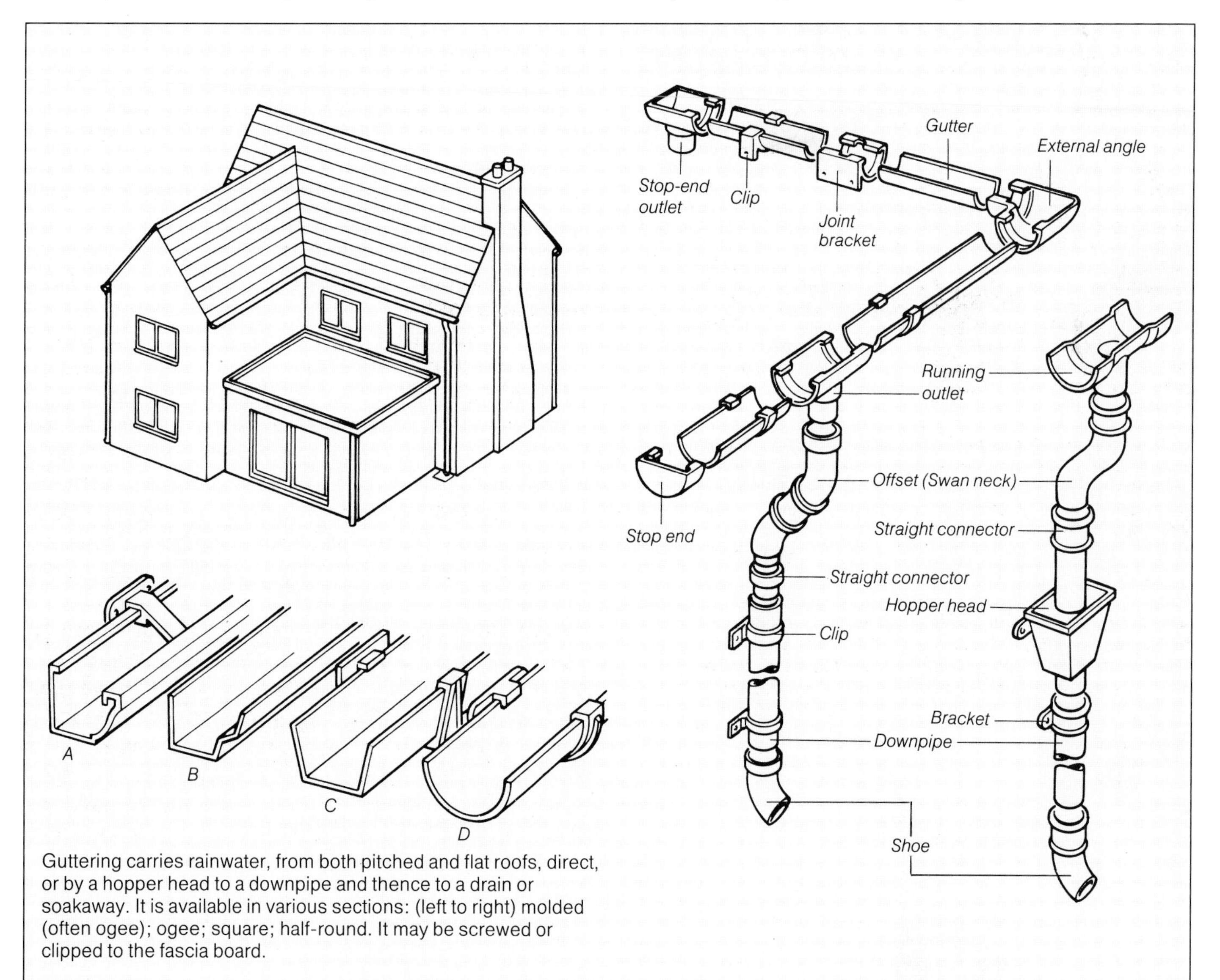

Guttering carries rainwater, from both pitched and flat roofs, direct, or by a hopper head to a downpipe and thence to a drain or soakaway. It is available in various sections: (left to right) molded (often ogee); ogee; square; half-round. It may be screwed or clipped to the fascia board.

Unfortunately, there are no adaptors available to join systems from different manufacturers, even though nominally they may have the same dimensions. For this reason you should always try to match existing plastic systems with components of the same make – and this is usually stamped on the gutters and fixings.

The other material used for rainwater systems is aluminum which, although more expensive than plastic, is just as easy to instal and can be painted, if preferred.

Potential problems

The most common problem with any rainwater system is a blockage, either in a gutter or downpipe. This will cause an overflow so the trouble is easily spotted. The remedy is simple: scoop out the debris and flush the gutter through, or push a wad of rag through the downpipe with a long rod to move the blockage. If this is not possible because the pipe incorporates an offset bend, you may have to dismantle it to reach the blockage. This should not be too difficult since the sections of pipe usually just slot into each other.

A sagging gutter may also cause an overflow and to cure this it might be necessary to reposition the brackets to ensure a steady fall towards the downpipe. Or you may need to reposition just one bracket.

FITTING A PLASTIC SYSTEM

If your existing rainwater system needs replacing, the easiest material to use is plastic. Get hold of a catalogue from your local supplier and inspect the existing system, noting the number and type of fittings you will need. Measure the length of the guttering and of the downpipe, allowing a bit extra for trimming.

You should also measure the size of the gutter between the rims and buy the equivalent size; in this way you will be sure that your new system will be able to cope with the amount of water coming off the roof. However, if you intend extending the system for any reason, it may be worth choosing the next size up.

Plastic and aluminum systems come in various brands, but there is little to choose between them, so stick to the one that is most readily available. The guttering, pipes and fittings are obtainable in gray, black or white, but they are easily painted if none of these colors fits in with your exterior color scheme.

Safe access is essential when you are replacing guttering and the best thing to do is to rent a scaffold tower. However, if you have no choice but to use a ladder, make sure its feet cannot possibly slip and that the top is tied to an eye screwed into the roof fascia board or house wall.

Removing the old system

The gutters may rest in brackets screwed to the fascia board or be screwed directly to the board. In the former case, lift out the sections of gutter and either unscrew or saw off the brackets.

The outlet section of gutter should simply lift out of the top of the downpipe. Then remove the downpipe; unscrew or saw through its mounting brackets and lift each section clear.

Essential preparation

Once you have removed all the old system, you can inspect the fascia board for signs of damage or rot. If it is in bad condition, it should be levered off and a new board nailed to the ends of the rafters. Otherwise, clean it off with a stiff brush, fill any old screw holes with a good quality exterior filler and paint it with primer undercoat and finish coat.

Similarly, brush down the wall behind the downpipe, fill the old pipe bracket screw holes and, if the rest of the wall is painted, treat it with the same color.

Installing the new guttering

The new guttering must have a steady and gradual fall towards the downpipe end if the water is to flow away efficiently. This fall should be about ¼in in every three feet and it is arranged by positioning the gutter brackets progressively lower along the fascia board.

Begin by screwing one bracket to the end of the fascia furthest from the downpipe and as close to the eaves as possible (bearing in mind that the gutter must be slipped into it). This bracket should be 6in in from the end of the fascia. Drive a nail into the fascia level with the top of the bracket and tie a length of string to it. Take the string to the other end of the fascia and pull it taught while someone else checks that it is horizontal with a spirit level. Lower the string enough to match the required fall and tie it to another nail hammered in place. Fit another bracket to the fascia at this point, aligning the top level with the string and 6in in from the end of the fascia.

Working towards the downpipe end start fitting the lengths of gutter into the brackets, spacing the latter at 3ft intervals and setting extra brackets on each side of any joints. Use galvanized screws to hold the

1 Levering away an old downpipe; if it is fixed securely, cut through the bolts with a hacksaw, close to the wall.
2 Lowering an old section of guttering on a rope.
3 Positioning the new outlet; fit it low down on the fascia board as the water must drain down to this point.
4 At the far end of the run, screwing the fixing bracket high up on the fascia to give a minimum fall of ¼in per 3ft.
5 Stretching a string line between the outlet and other bracket; check with a spirit level that you have the necessary fall.
6 With the string line still in place as a guide to height, fix the intermediate brackets at the recommended intervals along it.
7 Clip in the guttering along one run before fitting a corner piece, followed by the guttering on the other side of the corner.
8 Fitting a stop-end to the end of the guttering; if the outlet is not at one end of a run, fit a running outlet and two stop-ends.
9 Fitting a square-to-round adaptor to enable the run to be continued in guttering of a different section to the original one.

brackets in place and line up each with the string line.

The first length of gutter should have a stop end clipped to it to close it off; the end of the gutter should project 2in beyond the end of the fascia board.

The method used for joining lengths of guttering incorporates a rubber seal which is held to the plain end of the next length by a plastic spring clip. In other systems both ends are plain and a completely separate clip-on jointing piece is used. Sometimes this joint doubles as a support bracket.

You will have to trim at least one length to size. Measure for this very carefully, making sure it overlaps the rubber joint seals fully. In some cases, where the gutter is to run around a corner, it is easier to fix the corner piece in place first and then measure up to it for the gutter pieces on each side.

You can cut the guttering with a hacksaw, taking care to keep the cut square. Clean off any burrs with a file. Some types need notches cut in their edges to accommodate the securing clips and these can be made with a hacksaw and file, or you can hire a special tool at a relatively low cost.

When you reach the downpipe end, fit the outlet section and another stop end so that it projects 2in beyond the end of the fascia board.

Fitting the downpipe

If the downpipe needs an offset bend at the top to set it out from the wall so that it will connect with the gutter outlet, this should be made up first. Then you can offer it up to the gutter and determine the downpipe position. You may be able to buy a one-piece offset bend; if not, you can make one up from two offset connectors and a short length of straight pipe. In this case, the pieces must be solvent-welded together to ensure a watertight assembly. The top socket of the offset bend fits over the top of the gutter outlet.

1 Hanging a plumb-line from the outlet to mark the position of the leader.
2 Fitting a swan-neck to guttering on overhanging eaves; two-part swan-necks allow sideways adjustment of the leader position.
3 Marking the position on the wall of the top leader bracket; this should be a little distance below the outlet.
4 Fixing the bracket block of a two-part pipe bracket to the wall; this type permits all wall fixings to be made without the pipe in position.
5 Sliding the pipe clips onto the pipe to coincide with the bracket blocks; the wedge-shaped connector should taper downwards.
6 Screwing the pipe clip to the block after connecting to the swan-neck; the pipe socket, whether integral or separate, should be uppermost.
7 Connecting lengths of pipe; joints are not solvent-welded, and an expansion tolerance of ⅜in should be left between lengths.
8 Fitting the shoe to the bottom length of pipe; angle it away from the wall, 2in above the ground or splash block and solvent-weld it to the pipe.
9 Connecting the downpipe to a drain; fit a caulking bush into the drain socket, solvent-weld the pipe and mortar in place.

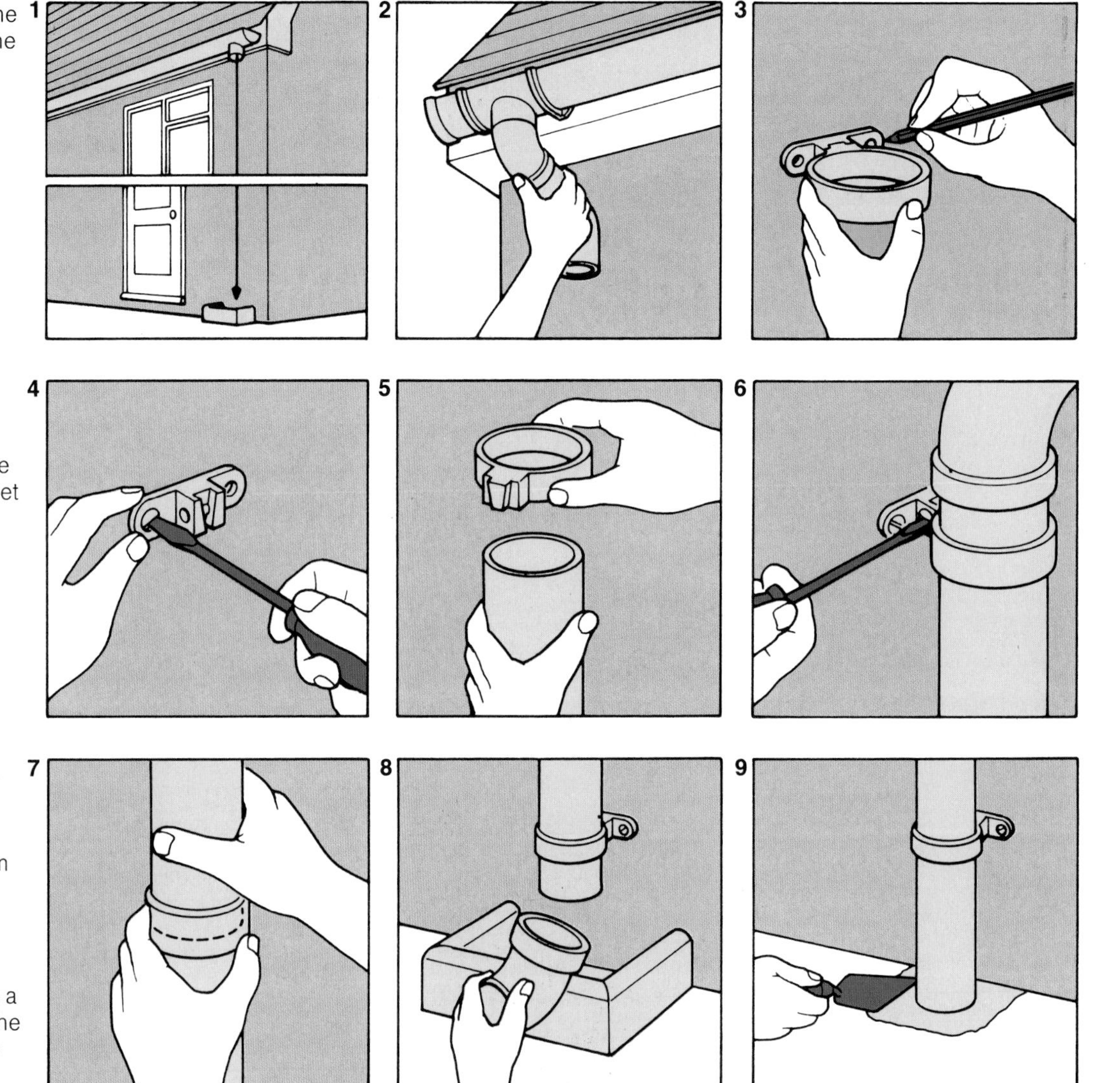

Use a plumbline to mark a vertical guideline on the wall for the pipe bracket positions. Then assemble the pipe, working downwards and screwing the brackets to plugged holes drilled in the wall. The bottom of each section pushes into the wider socket of the section below.

If the downpipe is to discharge over a gully, a "shoe" fitting is solvent-welded to the bottom of the pipe. This deflects the water away from the house wall. It should terminate about 2in above the gully grid to prevent water splashing up the wall. If a direct connection to an underground pipe is needed, the last section of pipe is solvent-welded into a socket which closes off the top of that pipe.

Finally, test the whole system by pouring water into the gutter making sure there are no leaking joints.

CHAPTER 4

INSTALLING DOMESTIC APPLIANCES

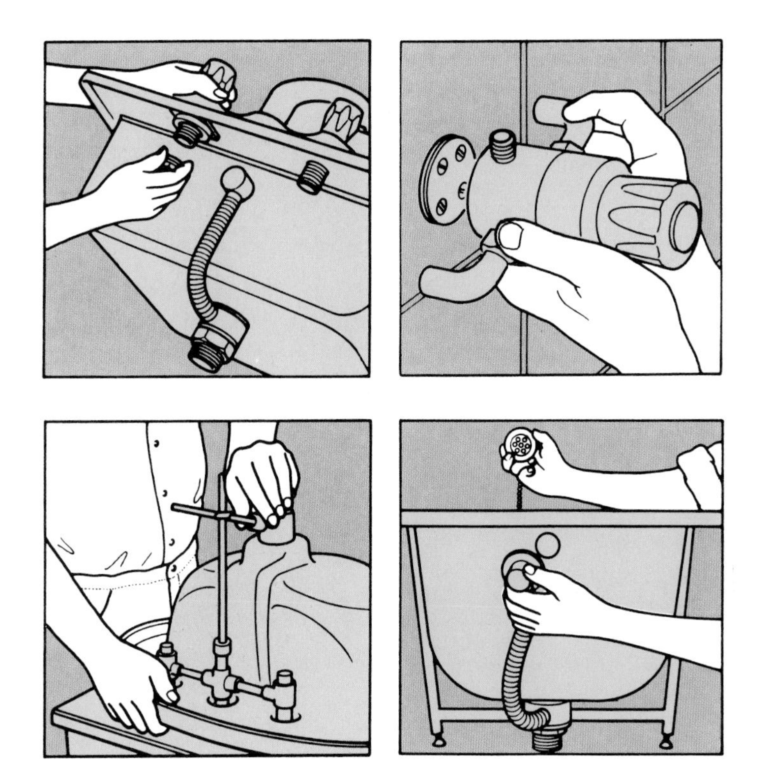

Having mastered the basic requirements given in the previous chapters — the ability to install pipework, make watertight joints and connect up to fittings — the home plumber has all the necessary skills to fit new appliances to his plumbing system.

Whether you wish to modify your existing plumbing or run new pipework to a different location, the following pages explain in detail how to install a variety of appliances ranging from a kitchen sink and garbage disposal unit to a hand basin, bathtub and shower, in fact, everything the home improver needs to maintain and upgrade his plumbing system.

INSTALLING A NEW SINK

The kitchen sink will probably be the first item you want to replace when modernizing a kitchen; there is now a very wide selection of kitchen sinks to choose from. They can be stainless steel, enamelled steel, ceramic or glass fiber and come in a wide range of colors to match any decorative scheme. You can buy single or double bowl sinks, sinks with single or twin drainers, individual bowls and drainers, bowls for waste disposal units, sinks that act as tops for kitchen units and others for setting into worktops. All can be installed in the same basic manner, with slight alterations to the way they are mounted, the faucets fitted and wastes installed.

It is important for the sink to be out of commission for the least amount of time possible during the alteration work. To ensure this, assemble the new sink with its faucets and waste outlet before removing the old one. Likewise assemble any base unit but do not fit the sink top to it until it is in position. If you are fitting a new worktop to an old sink base unit for an inset sink, you can make the cutout for the sink before fitting the top. Use a powered jig saw or saber saw for the job.

Assemble the waste outlet to the new sink first, bedding it on a layer of plumber's putty if no rubber or plastic gasket is provided by the manufacturer. Some sinks may have an overflow duct formed integrally, but others will have a separate overflow assembly. In the former case make sure the slot in the waste outlet is aligned with the duct. In the latter, the overflow pipe has a "banjo" fitting at its end which fits over the outlet tail. Make sure the hole in this is aligned with the slot in the outlet before securing it with the backnut on the outlet.

Fit the faucets in place, applying a layer of plumber's putty round the base of each one if no gasket is provided. If you decide on a mixer valve, make sure it is for kitchen use only. These do not allow cold mains water to mix with stored hot water within the body of the faucet, avoiding the possibility of contaminating the mains supply.

It may be necessary to fit "top hat" spacer washers to the faucet tails before the backnuts can be tightened fully. To aid connection to the supply, fit

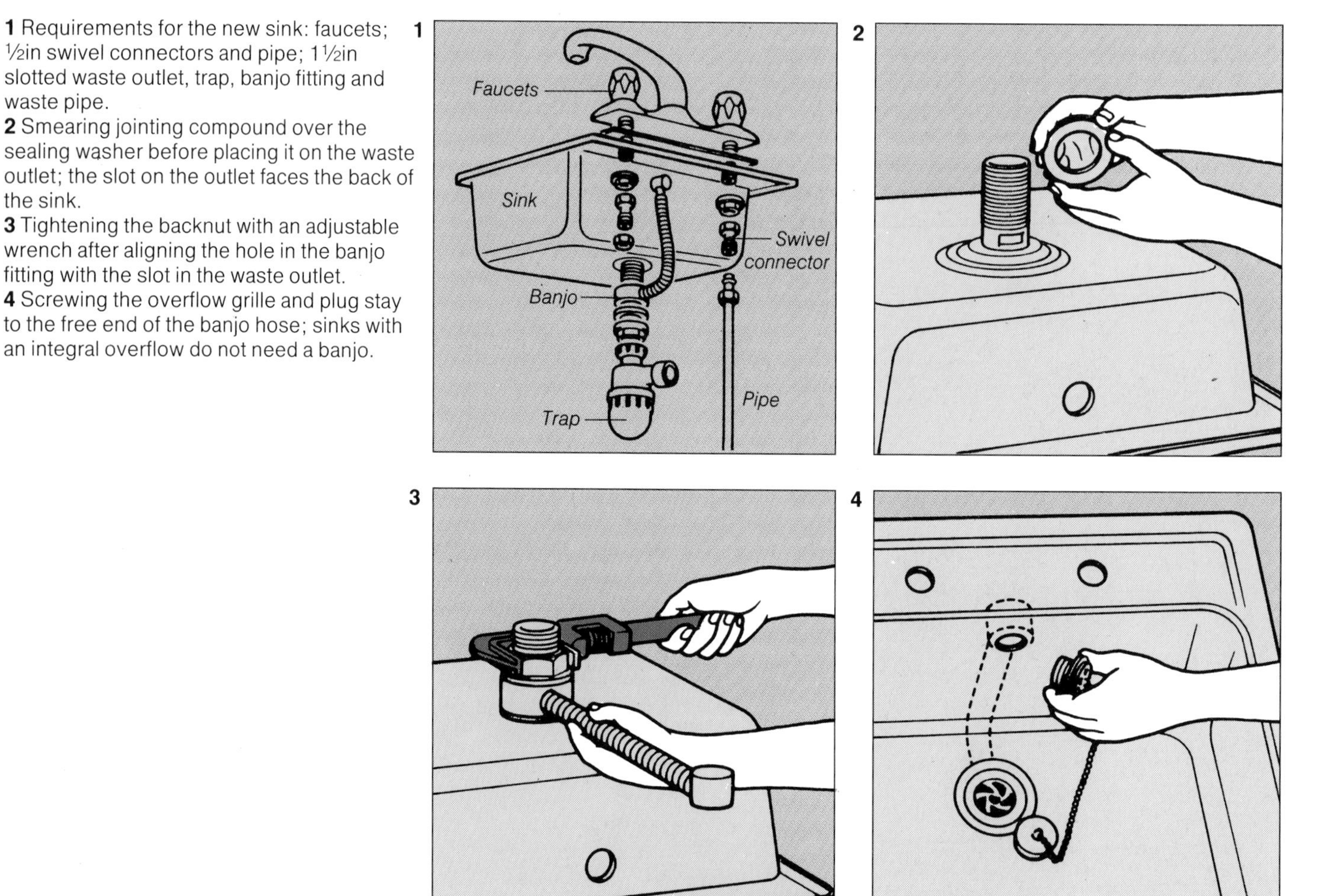

1 Requirements for the new sink: faucets; ½in swivel connectors and pipe; 1½in slotted waste outlet, trap, banjo fitting and waste pipe.
2 Smearing jointing compound over the sealing washer before placing it on the waste outlet; the slot on the outlet faces the back of the sink.
3 Tightening the backnut with an adjustable wrench after aligning the hole in the banjo fitting with the slot in the waste outlet.
4 Screwing the overflow grille and plug stay to the free end of the banjo hose; sinks with an integral overflow do not need a banjo.

lengths of flexible copper pipe with a swivel connector at one end to each faucet tail. These should be ½in diameter.

Removing the old sink

First turn off the water supply to the hot and cold faucets and open the faucets to drain the pipes.

You may find difficulty in reaching the faucet tails to disconnect them and it will be easier to cut through the supply pipes as close to the faucets as possible. If the pipes are buried in the wall, chop out the plaster around them to make the cuts. Unscrew the waste trap from the outlet or disconnect it from the waste pipe.

Break any seal between the sink and wall with a hammer and chisel and lift the sink away if it rests on metal brackets. Alternatively unscrew it from its base unit and lift it out. Remove any brackets with a hacksaw.

Installing the new sink

Move any base unit into position and set the sink top in place. Mark and cut back the supply pipes so that the flexible pipes can be connected without too sharp a bend. You can use either compression or capillary joints to make the connections, but some Building Codes may insist on capillary joints for any pipes carrying mains water.

Screw on the trap, either using the original or a new trap depending on whether the waste pipe is horizontal or vertical. It must have a 3in deep water seal for best results.

If the original waste pipe will not line up with the new sink, or if it is made from lead, then it would be better to install a new plastic pipe of 1½in diameter, making sure it has a fall of ¾–1¾in per yard towards the drainage point. Use either push-fit connectors or solvent-weld fittings.

The waste pipe should discharge into a soil stack; wherever possible connect up to the original pipe. If you have to cut a hole through the wall for the pipe, you can rent a masonry core drill to cut it in one operation. It is important to seal around the pipe with a non-setting caulking to make the joining completely watertight, and then restore the water supplies to both faucets.

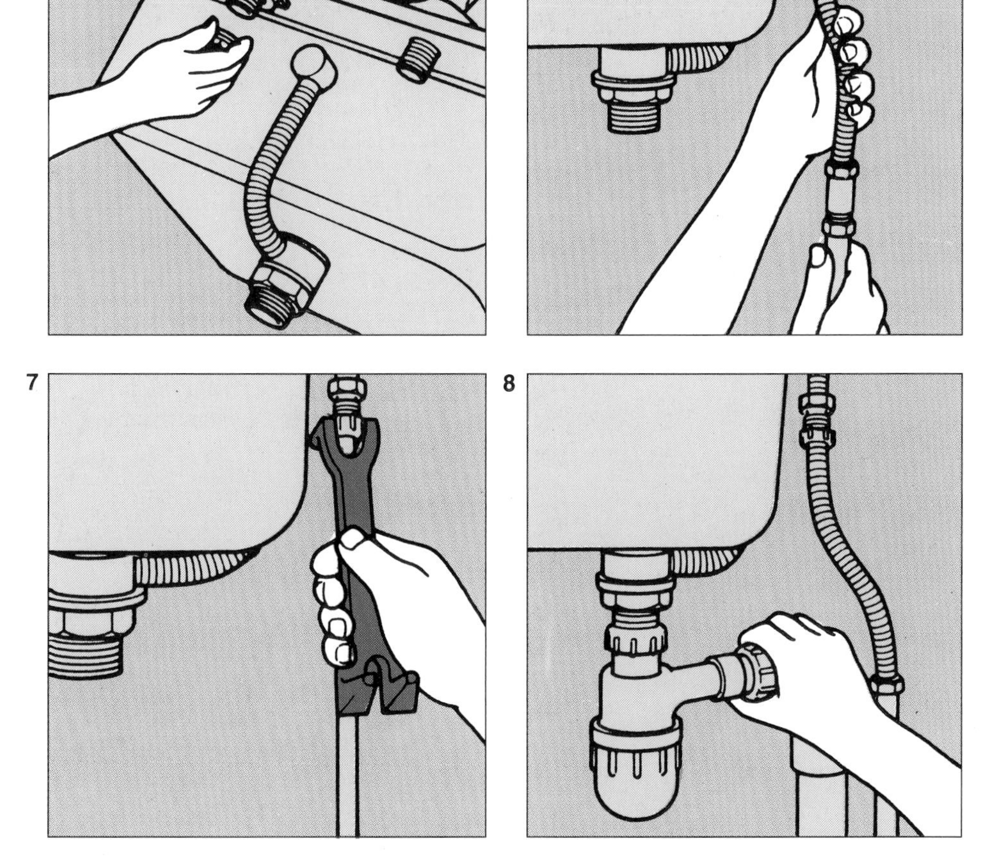

5 Securing faucets with backnuts after sliding on top-hat washers; position any sealing gasket between the faucet body and the sink top.
6 Using a corrugated faucet connector for ease of bending; connect it first to the faucet tail, then to the end of the supply pipe.
7 Tightening the compression joint with a basin wrench; this allows access to the right corners often encountered behind sinks.
8 Connecting the trap to the waste pipe after screwing it to the outlet; double sinks should have a fall from each trap before teeing together.

• CHECKPOINT •

Fitting a waste disposer

A waste disposal unit is a useful appliance to have in a kitchen; it contains a set of hardened steel blades driven by a powerful electric motor which will grind up most forms of kitchen waste. The ground waste is mixed with water inside the unit to make a slurry which can be flushed down the drain. The waste disposal unit is suspended directly below the sink, fitting between the waste outlet and the trap. Some sink units have a small separate bowl specifically for waste disposal.

Practically all waste disposal units need a larger than normal waste outlet in the bottom of the sink – about 3 ½in in diameter – and this may mean buying a new sink with this size outlet. However, some stainless steel sinks can have their outlets enlarged by a special hole saw which you may be able to rent. Ceramic or enamelled sinks cannot be cut and must be replaced. Another point to consider if you are thinking about enlarging the existing outlet hole is whether you will still be left with a circular depression large enough for the outlet flange to be inbedded; it must be below the level of the sink bottom otherwise the sink will not drain properly.

After installing the new larger outlet fitting in the sink, the waste disposal unit grinding compartment is mounted beneath it by an assembly of circular plates which ensure a watertight seal with the outlet, and support the weight of the unit. An angled outlet pipe is attached to the side of the grinding compartment and a normal P or S trap connected to this. A bottle trap should not be used since it takes too long to discharge its contents.

As with other kitchen sinks, the waste pipe should be of 1½in diameter and it should have a fall of about 15° to the drainage point.

Once the waste pipe has been run in, the motor compartment can be fitted beneath the grinding compartment and the wiring connections made in accordance with the manufacturer's instructions.

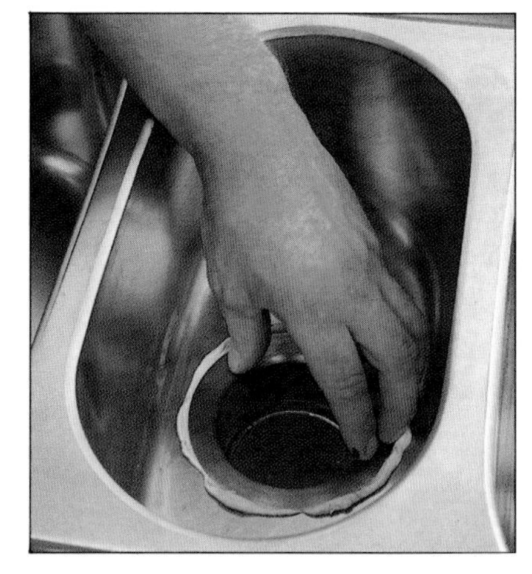

Bedding the 3½in waste outlet into the sink on a ring of plumbers' putty; some units will fit on a standard waste.

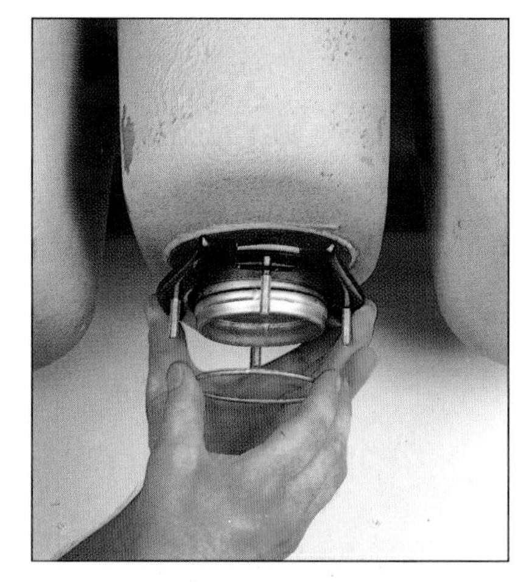

Assembling the suspension; top to bottom: rubber washer, pressure plate, suspension plate and circlip.

Tightening the Allen screws with the key provided; locking the suspension plate against the pressure plate.

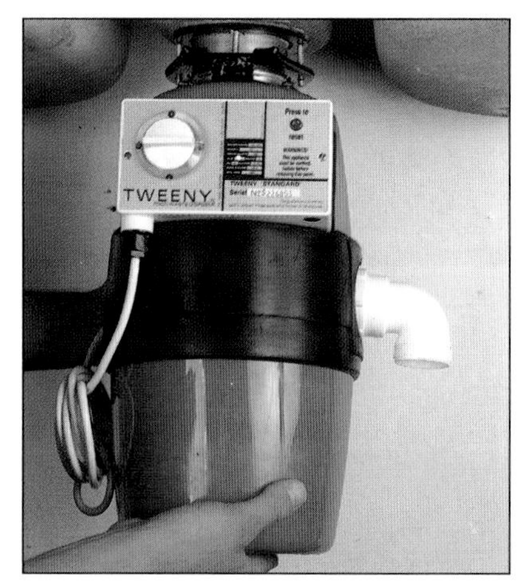

Attaching the waste disposer to the suspension after fitting the waste elbow; hinges forwards for servicing.

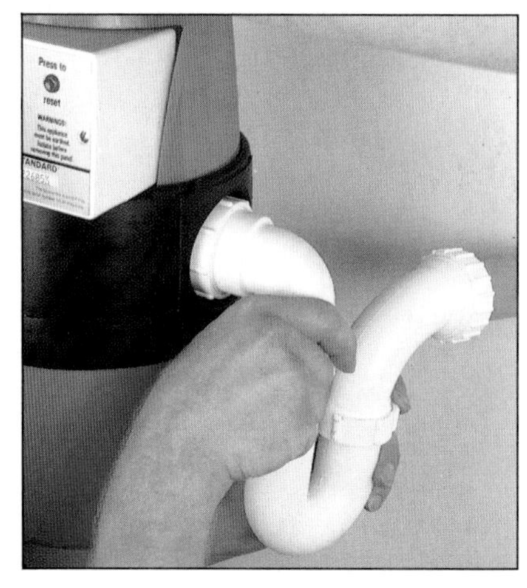

Tightening the swivel connections on the trap after connecting to the elbow and waste pipe; do not use a bottle trap.

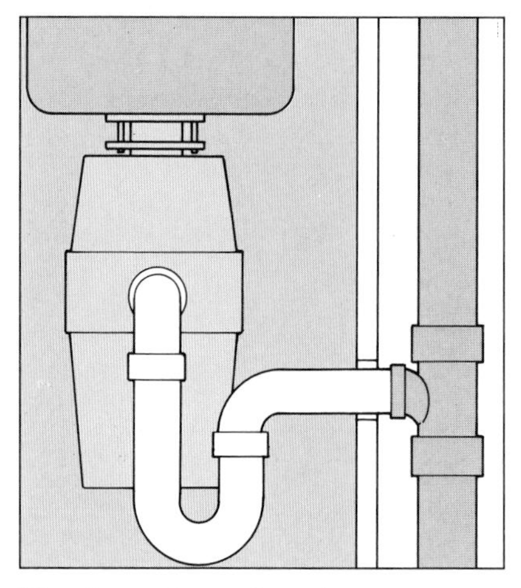

The disposal must have its own connection to the stack and not be interconnected with the sink drain.

• CHECKPOINT •

Using washing machine and dishwasher kits

Making permanent connections to the supply and waste pipes is a much neater and more efficient way of installing a washing machine or dishwasher than pushing the machine's hoses onto the kitchen faucets and hooking the waste hose over the edge of the sink. There are now plumbing kits available that make this job extremely simple.

Some machines need both a hot and cold water supply whereas others may need only cold water. It does not matter which, the method of connection remains the same. It is fairly common to take the cold supply from the kitchen cold faucet supply pipe, but first check the instructions supplied with the machine as it may not always be suitable. The manufacturer's instructions will also state what water pressure is required. If you intend to install a machine upstairs, first make sure that there is sufficient pressure for it to function properly. To install a machine downstairs, the water pressure is not usually a problem, if you can take it from the main supply to the kitchen sink – but check with your local Plumbing Code before connecting more than one machine.

Water supply

To supply the machine with water you simply break into the nearest supply pipes with compression or capillary T fittings, or use an automatic connector such as that supplied with a plumbing-in kit. This allows the connection to be made without draining the pipes. Then run lengths of ½in pipe to where the machine is positioned. Each pipe should terminate in a combined stop-valve and washing machine hose connector. Screw the hoses to the connectors to complete the supply side.

Waste

For the waste you can use a conventional washing machine standpipe kit comprising a vertical length of 1½in waste pipe with an integral P trap at the bottom. This should be connected to a waste pipe that passes close by or is connected to the kitchen sink waste pipe with a T fitting. Then hook the waste hose into the top of the standpipe.

Alternatively, you can use a washing machine waste pipe connector. This clamps around the existing kitchen sink waste pipe and a tool supplied with the fitting is used to bore a hole in the pipe. A hose connector is then fitted over the hole and the machine's waste pipe pushed on.

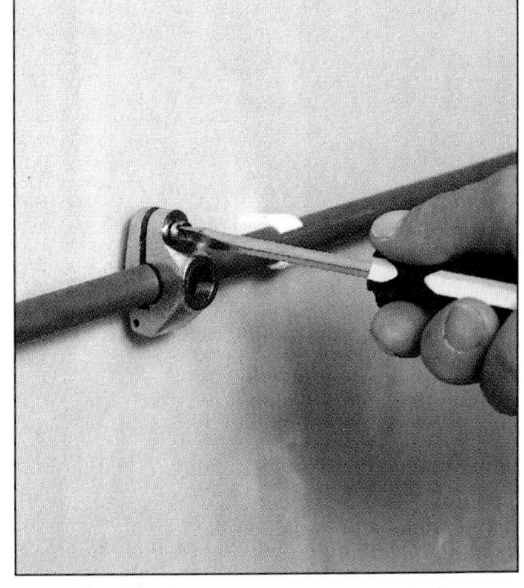

Clamping the baseplate of a plumbing-in kit to the supply pipe; the baseplate hinges open to pass round the pipe.

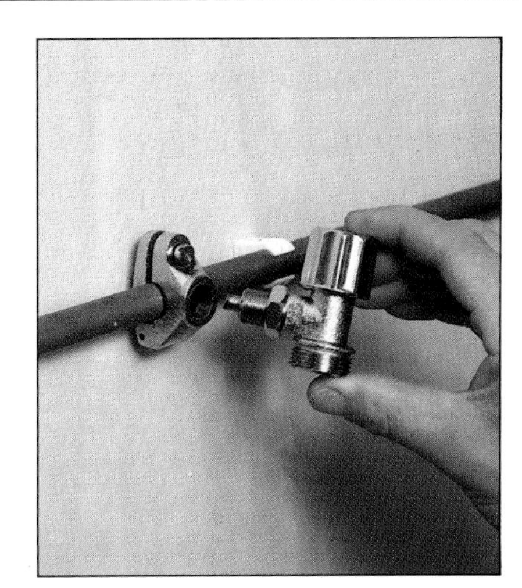

Screwing the stop-valve body into the baseplate; a cutter bores into the pipe as the stop-valve is screwed in.

Attaching a conventional stopcock to a branch pipe after breaking into the supply with a tee fitting.

Attaching the trap to a washing-machine standpipe; the hose from the machine hooks into the top.

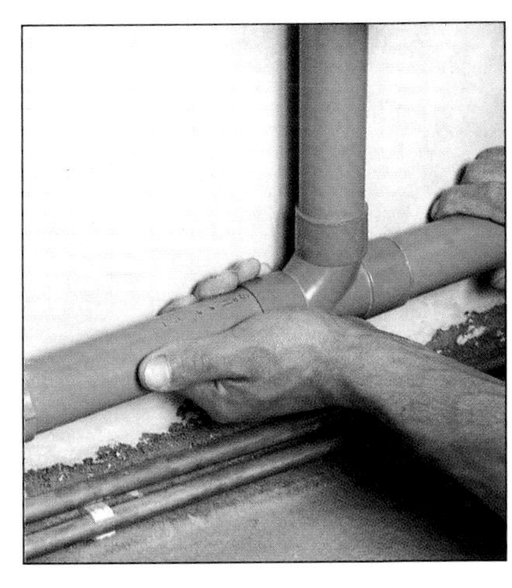

Connecting into the kitchen sink waste pipe using a sanitary-tee; the curve should point towards the stack.

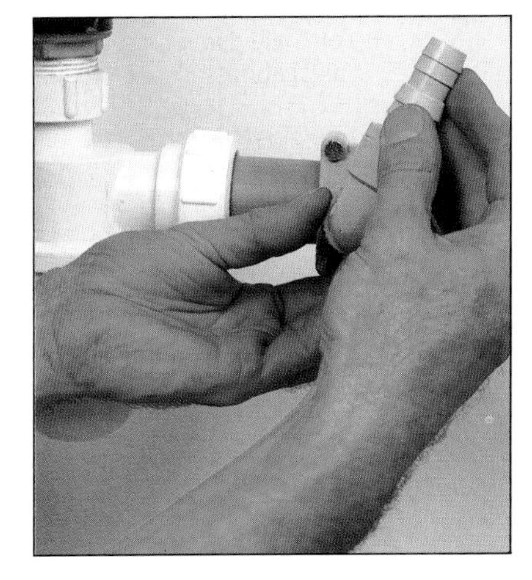

Screwing a waste connector into the strap clamped around the sink waste pipe; a tool is supplied to bore into the pipe.

FITTING A NEW HAND BASIN

If you are updating an old bathroom or remodeling it, one of the jobs you are likely to do is replace the hand basin. Modern basins may be made in ceramic, enamelled steel or glass-reinforced plastic. They come in a variety of shapes and colors, usually designed to match with a range of other bathroom suites. Basins may be wall-mounted, recessed into the wall, let into flat surfaces or mounted on pedestals; even double basin units are made. Wall-hung basins come with supporting angle brackets. Generally, the device will be centred, then leveled over the drainpipe at 31 to 38in above the floor.

Arrange this so that the basin is out of action for the minimum amount of time. To do this you will find it easier to prepare the new basin by fitting its faucets and waste outlet before removing the old basin. If necessary, use top-hat spacers on the faucet tails to allow the backnuts to be tightened fully and bed the outlet on plumber's putty or the gasket provided. To aid connection to the supply pipes, attach lengths of flexible pipe to the faucet tails.

Removing the old basin

Turn off the water supplies to the faucets and open them to drain the pipes (see page 13). Unscrew the faucet connectors and waste trap, having a bowl ready to catch any water that may spill out. If you cannot undo the faucet connectors, cut through the pipes just below them. If you are moving the position of the basin, you may prefer to cut the pipes below floor level and make up new extensions to the new position. If the pipes are lead, you should replace them completely.

With the pipes disconnected, unscrew the basin from the wall and lift it away, then remove the brackets or pedestal.

Note: You may find it easier to suspend the weight of the basin from above or find a helper to support it while you remove the last lug before lifting it out.

Installing the new basin

Mark, drill and plug any wall fixing holes for the new basin and its brackets, making sure the basin is at a comfortable height, (of course, the height is predetermined with a pedestal basin). Fix the basin to the

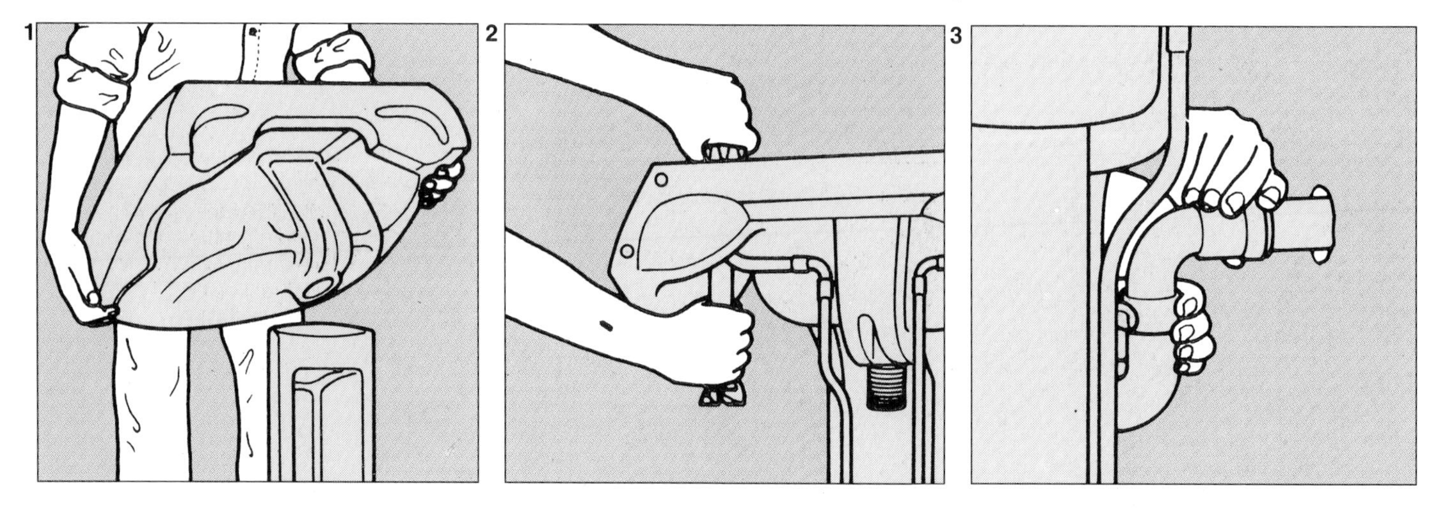

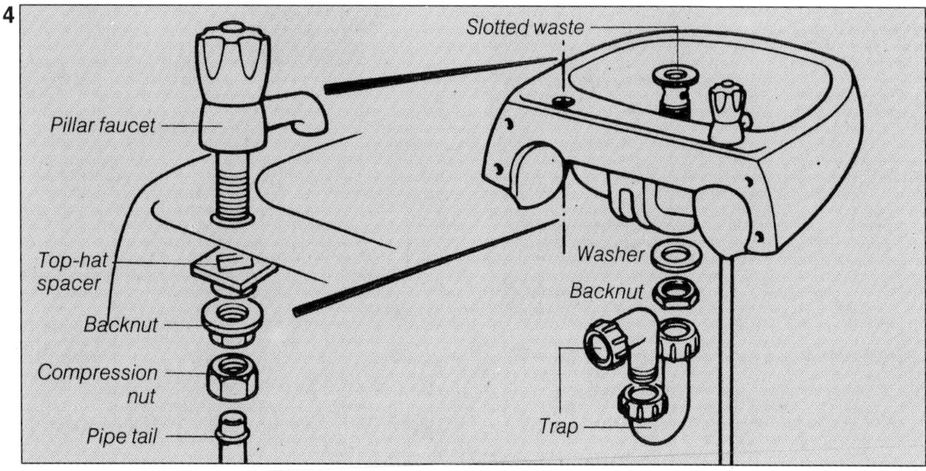

1 Lifting a hand basin onto its pedestal to measure the length of the supply pipes and the height of the waste outlet and fixing holes.
2 Connecting the supply pipes to the faucet tails using a basin wrench; bend or joint the pipes to run neatly within the pedestal.
3 Connecting the trap to the waste pipe after moving the basin and pedestal into position and securing the basin to the wall.
4 Components needed to plumb in a hand basin: faucets, connectors and pipe; ¼in waste outlet, trap and waste pipe.

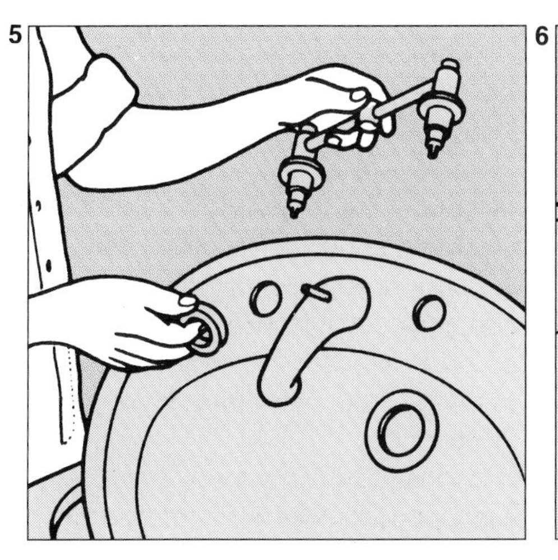

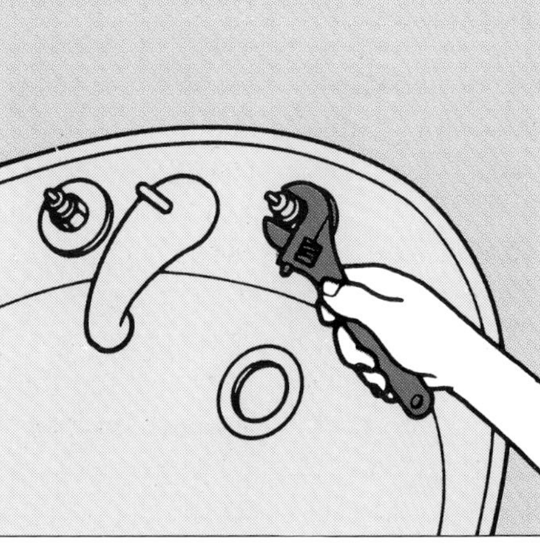

5 Fitting a vanity unit. After fixing the outlet spout; checking that the connecting pipes on the water inlet are the correct length.
6 Fixing the water inlet assembly to the unit, ensure that all washers supplied are fitted as instructed, or the inlet will not be secure.

7 Attaching the pop-up waste operating rod to the underside of the water inlet; adjust the length of the rod according to the depth of the unit.
8 Applying a gasket to the opening in the base unit before inserting the basin, to ensure a watertight seal.

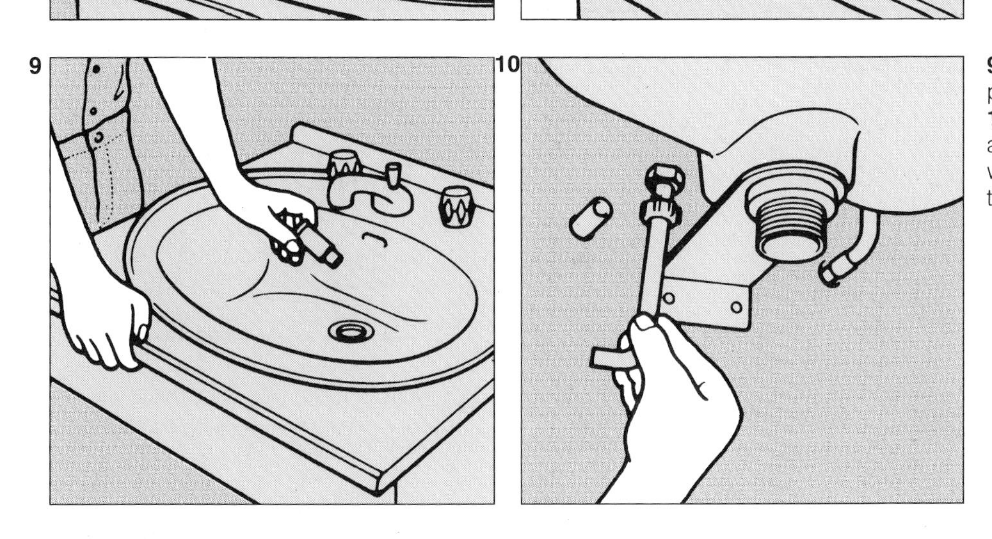

9 Attaching the pop-up waste plug after pressing the basin gently onto the gasket.
10 Connecting the supply pipes after attaching the angle bracket, between washers, to the waste outlet and screwing it to the wall.

wall and cut back the supply pipes so that the flexible copper pipes can be connected to them with a smooth radius. It is a good idea to install front legs on the basin for extra stability.

If the basin is pedestal-mounted, you can fit much longer pipes to the faucets before positioning the basin, running them down the inside of the pedestal for concealment and connecting them to the supply pipes near floor level.

If you are lucky, the original trap can be screwed to the outlet of the new basin, however it may not align exactly in which case you could fit a telescopic version. If you are completely renewing the waste run, use a P or bottle trap and $1\frac{1}{4}$in waste pipe (unless the waste run is over 5 feet long, in which case it should be $1\frac{1}{2}$in pipe). Note, when the pipe is being installed the pipework should have a gradual fall and discharge into a soil stack.

REPLACING AN OLD BATHTUB

Replacing an old bath with a new one is carried out in much the same way as fitting a new hand basin; connections have to be made to the hot and cold water supply pipes and to the waste pipe, or new pipes installed (particularly if the old ones are lead). You will probably find that the connections at the faucets and waste outlet are difficult to reach, and this may mean some careful preplanning before sliding the new bath into place.

Choosing a new bath

Gone are the days of the old cast iron bath with white enamel finish; baths nowadays come in all shapes, sizes and colors. The oblong bath is still most popular, but you can also buy triangular baths for fitting in corners, oval baths and short, deep tubs for use where space is limited. Faucets can often be fitted at the side of the bath or at a corner, as well as in the traditional end position.

Although you can buy enamelled, pressed steel baths, the most popular type now (particularly for handyman-installation due to their light weight) are plastic baths. These usually have a non-slip finish molded into the bottom and come with a steel and wood supporting cradle with attachments for pre-molded trim panels.

One point to watch with a plastic bath is that you cannot make soldered capillary joints anywhere near it because there is a danger of the blowtorch melting the plastic.

Removing the old bath

Cut off the water supplies and remove the trim panels. If you can reach the faucet tails, disconnect the supply pipes from them; if not cut through the pipes as close to the faucets as possible. Treat the waste pipe in the same way and also cut through any overflow pipe which will pass out through the wall. The latter will not be needed any more since all modern baths have an over-flow assembly similar to that used on a basin.

Carefully break any seal between the bath and wall with a cold chisel and pull the bath away.

Because a cast-iron bath will be too heavy to

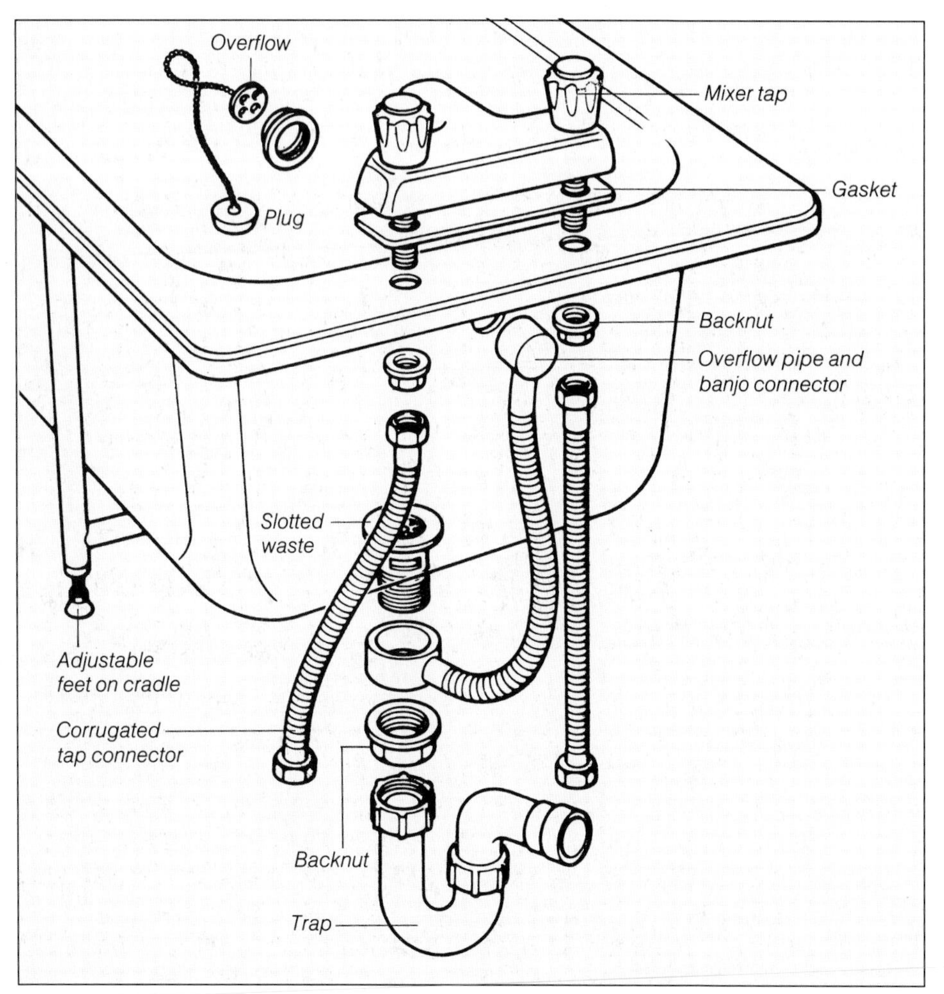

Components needed to plumb in a bath tub: faucets; ¾in corrugated pipes; 1½in waste outlet, P-trap and banjo fitting with overflow pipe. You will need a slotted waste outlet unless the overflow is attached directly to the trap; a shallow trap may be needed if height is limited.

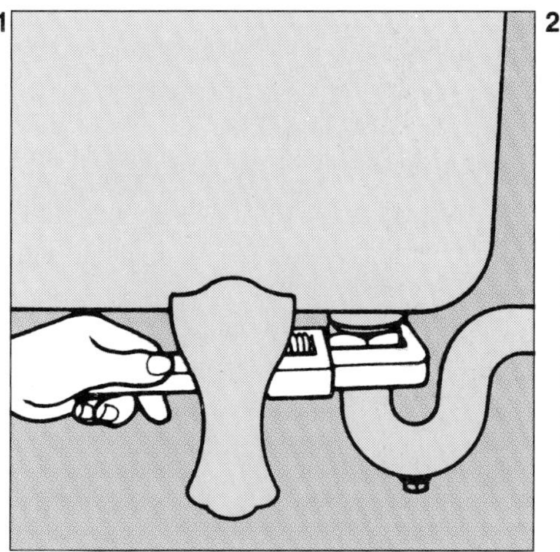

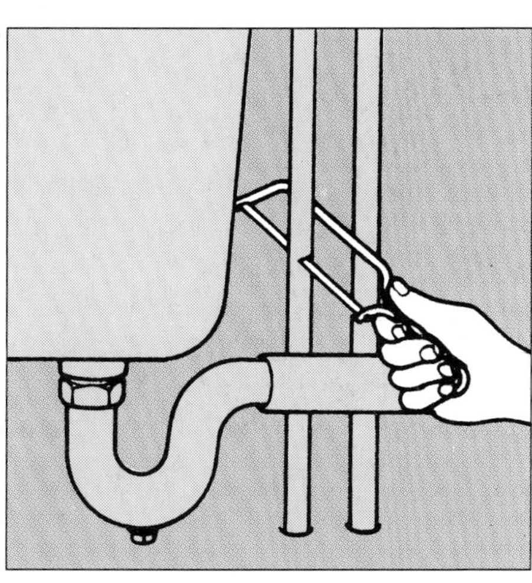

1 Removing the old bath by first disconnecting the trap using an adjustable wrench.
2 Cutting through the supply pipes with a hacksaw; leave sufficient length to attach flexible pipes.

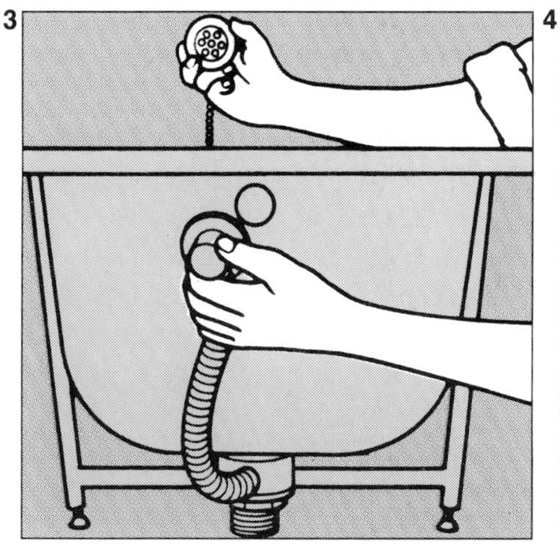

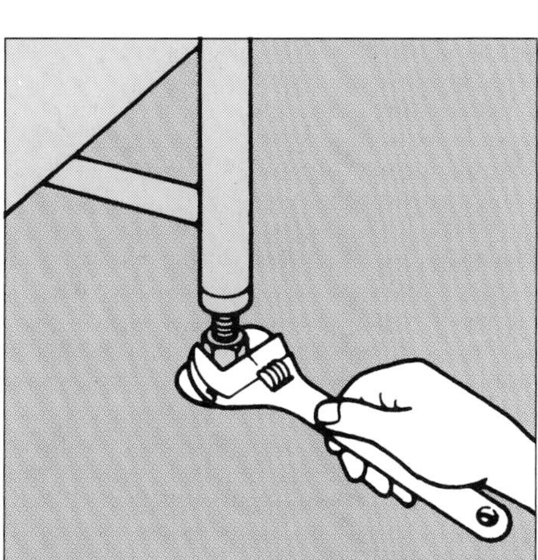

3 After fitting the waste outlet and banjo to the new bath, connect overflow pipe to plug holder; remember to fit the sealing washers.
4 Adjusting the leveling feet; the bath should be level crossways and lengthways, as a drainage fall is built into the bottom.

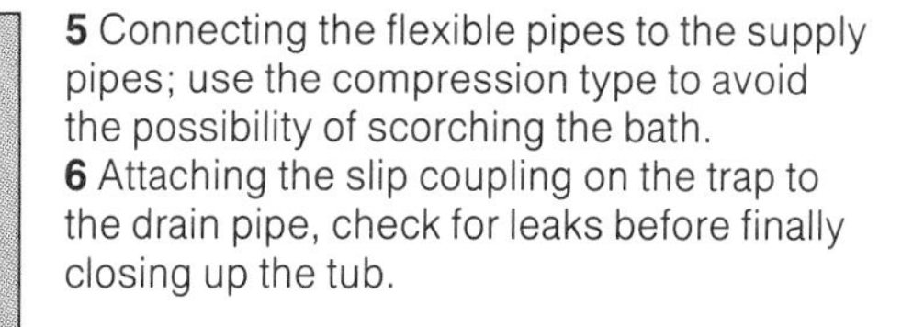

5 Connecting the flexible pipes to the supply pipes; use the compression type to avoid the possibility of scorching the bath.
6 Attaching the slip coupling on the trap to the drain pipe, check for leaks before finally closing up the tub.

manhandle down stairs, break it into smaller pieces with a sledge-hammer after first draping a thick blanket or tarpaulin over it to prevent splinters flying about and causing untold damage.

Fitting the new bath

Assemble the faucets, waste outlet and trap to the bath in the same manner as fitting out a basin. Because of the lack of space below the faucet, fit lengths of flexible pipe to the faucet tails before locating the bath. Note that the pipes should be ¾in size.

Fit the supporting cradle and move the bath into position, making sure it is level from end to end and side to side. Connect the flexible pipes to the supply pipes using compression joints, and the trap to the waste pipe.

Finally, turn on the water and check for leaks before installing the side and end trim panels.

FITTING A LOW LEVEL TOILET

When modernizing a bathroom, you will want to get rid of an old toilet with a high level tank. Made of cast iron, these tanks tend to be noisy in use, prone to corrosion and unattractive to say the least.

Modern efficiency

Modern tanks have much more efficient and quieter flushing mechanisms than their high level predecessors; and toilet designs are better. Older examples simply relied on the force of water pouring in from the tank to carry away waste, whereas the latest versions incorporate valves which produce a strong siphonic action to suck the contents through the trap aided by the flow of water from the tank.

There are several different toilet/tank layouts to choose from: the tank can be wall-mounted and connected to the pan by a short vertical flush pipe, it can be concealed within a partition so that only the flush handle shows, or it can be mounted directly to the back of the pan (known as a close-coupled suite).

If you want to change a high level tank to a low one, you can get a special slim "flush panel" that will fit to the wall behind the pan and still allow the seat and lid to be lifted – a conventional low level tank would project too far out from the wall for this.

Removing the old tank and pan

Turn off the water supply, flush the tank to empty it and disconnect the pipework, cutting it if needed. The flush pipe should simply pull from the back of the pan.

Unscrew the tank from the wall and lift it from its brackets; take care since it will be very heavy.

The pan may incorporate a P trap connected to a waste pipe passing through the wall or an S trap connected to a pipe rising vertically from the floor. It may be screwed or cemented in place. Carefully break the seal with the pipe and remove the pan, clearing any old mortar from the floor. Then carefully chisel out the mortar from the pipe socket.

Fitting the new pan and tank

Connect the pan to the waste pipe with a wax ring connector and screw it to the floor using brass screws with lead washers under their heads. Do not cement the pan in place as it may crack as the mortar dries.

Use the flush pipe to determine the tank position and screw its brackets to the wall. Check that the tank is level, packing it out beneath if necessary, and screw it to the wall.

Connect the flush pipe and assemble the flushing mechanism according to the instructions supplied.

Either make up a ½in pipe extension from the old supply pipe to the tank, or run in a new pipe from a more convenient source. Fitting a stop-valve just before the cistern will make maintenance easier. You will find that most modern tanks and flushing services include an integral overflow pipe. Finally, turn on the water.

1 Disconnecting the supply pipe to the tank after turning off the supply and flushing the tank; some water will still be inside.
2 Loosen and remove the bolts holding the bowl to the floor, rock the bowl to loosen the seal and then lift the bowl away from the floor.
3 Clean off the old gasket material and install new flange bolts into the floor flange.
4 Fit new wax gasket to new bowl and lower bowl onto floor flange, press down to compress the sealing gasket and then tighten flange bolts.
5 With a separate tank, after mounting it on brackets, connecting the flush pipe; the other end fits into a rubber boot on the pan.
6 Attaching the water supply pipe to the ball-valve inlet; assemble the flushing mechanism following the manufacturer's instructions.

• CHECKPOINT •

Fitting a remote anxiliary toilet

It can often be useful to install an extra toilet – as part of an attic conversion, room extension or alteration to a basement, for example, or if you have an elderly or infirm person living with you. Unfortunately, you are often limited in your choice of sites for the appliance because of its need for a bulky 4in or 3in soil pipe. The position of the existing soil stack or underground drain effectively dictates where you can put a new toilet, unless you go to the expense of installing complete new drainage runs.

Freestanding pump and shredder unit

Fortunately, there is a piece of equipment on the market that can overcome these problems. It is a freestanding electrically-driven pump and shredder unit, which fits behind the toilet pan and is connected to its outlet by a rubber sealing collar. When the toilet is flushed the waste flows into the unit where a pressure switch turns on a set of grinding blades which reduce all solid waste to a slurry and switches itself off about 18 seconds later. This is then pumped through narrow bore pipes to a soil stack up to 97ft 6in away. It will even pump vertically up to a height of 6ft 6in. It is also usual for WC waste to be connected to the stack at least 8in above other waste connections.

The waste pipe can be of copper with soldered capillary joints or plastic with solvent-weld joints and needs a minimum bore of ¾in. Being so much smaller than conventional soil pipe, it can be run under floors, through partitions and along baseboards to make an inconspicuous installation. The unit also has a vent pipe which must be taken outside, and this should be considered when making a final decision on positioning.

The device is electrically operated and needs its own fused supply. If you install this type of WC in a bathroom, remember the flex outlet should be suitable for a bathroom.

Because the unit is rather unconventional, you must have the approval of your local Building Department before going ahead with the installation; some may not allow it, so check with them before you buy.

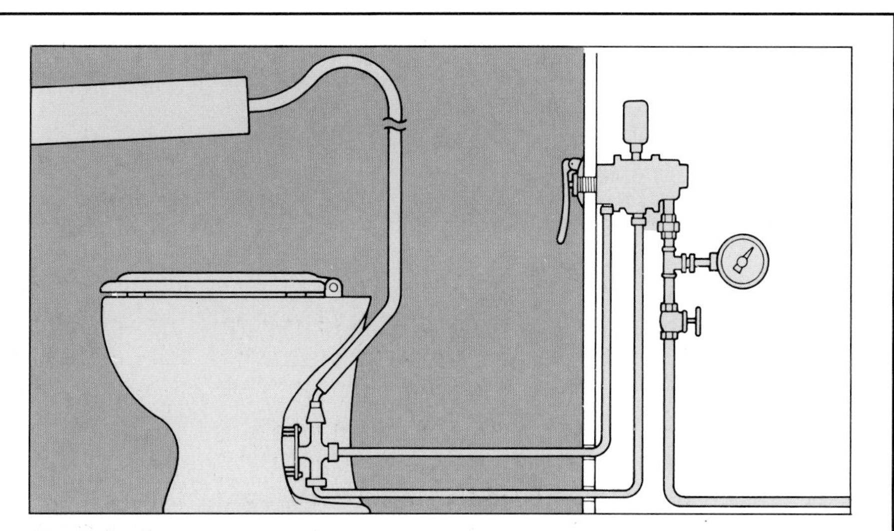

Water enters the flushing system through a double acting flush valve. Upon flushing, water is sent to the unit first by a small pipe connected to a disintegrating jet which breaks up the contents of the bowl. The valve then switches to a flushing jet which forces the contents of the bowl upwards to the soil stack.

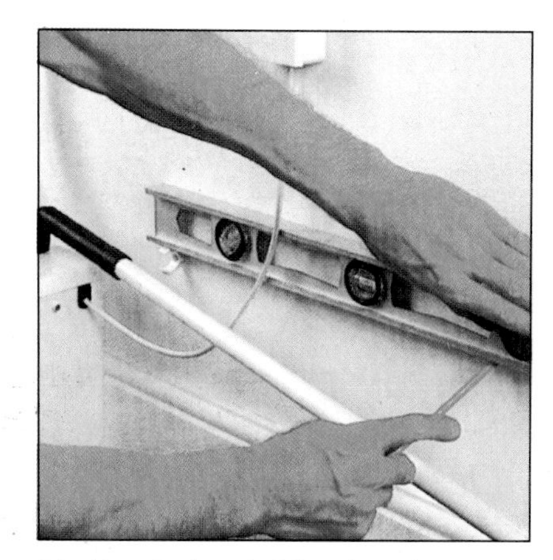

Marking the level of the pipe clips on the wall to give a fall of 1 in 200; clip regularly to avoid sagging.

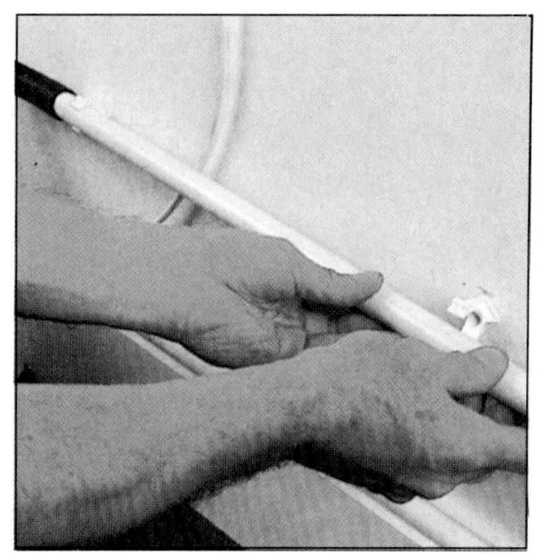

Pushing the pipe into the clips; fix all the clips to the wall before inserting the pipe.

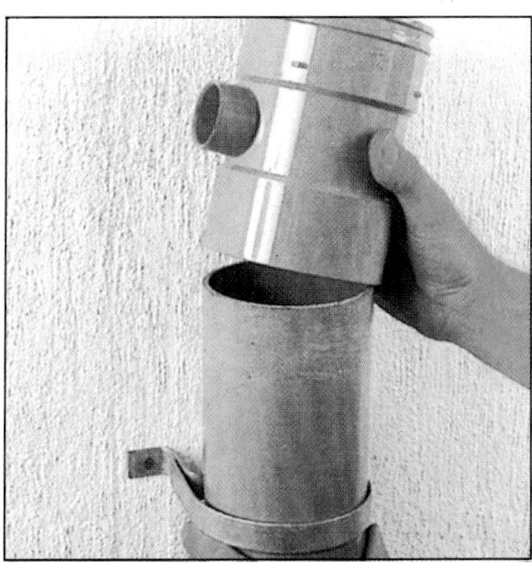

After cutting into the waste stack, solvent-welding a 1¼in boss outlet to it, to take the waste pipe.

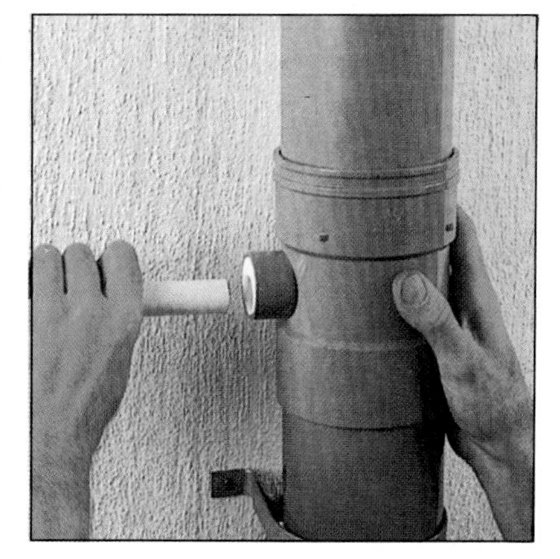

Connecting the waste pipe to the stack; a 1¼in to ¾in reducer is solvent-welded into the boss to accept the pipe.

INSTALLING A BIDET

Considered by many to be a luxury, a bidet is nevertheless a useful appliance to have in a modern bathroom. Those available come in shapes and colors to match other bathroom fittings so a co-ordinated look can be maintained.

There are two types of bidet: the straightforward over-rim bidet and the more complicated rim-supply and ascending spray types.

The over-rim bidet is very similar to a normal wash basin in that it is fitted with individual pillar faucets or a mixer, the spout of which discharges over the rim and into the bowl of the bidet. It is also plumbed-in in the same way as a basin.

The rim-supply and ascending spray type is rather more complex. The controls allow hot water to flow first through the hollow rim in order to warm it and make it more comfortable to sit on; the controls are then used to divert the flow to a vertical spray in the base of the bowl.

The spray outlet will be covered with water when in use and to prevent the likelihood of the water supplies being contaminated by back-siphonage, the feed pipes must be taken directly from the cold water mains supply and the hot water cylinder just as they would with a shower, but via a vacuum breaker device that is designed especially to prevent back siphonage and possible contamination of the drinking water supply.

Both types normally have a pop-up waste fitting which should be connected to a waste pipe and P-trap if the waste run is no longer than 5ft 6in or 2in pipe if it is longer than that, up to a maximum of 7ft 6in. The waste pipe can run directly to a soil stack. A 3in deep seal trap is needed.

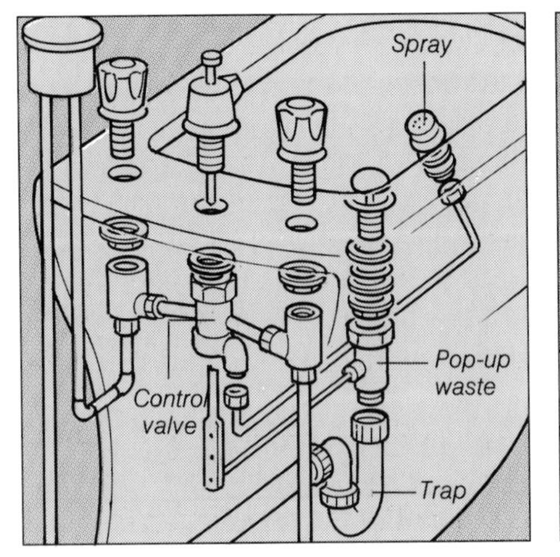

To connect ascending spray bidet: faucets; spray; vacuum breaker valve; control valve mixer; 1¼in pop up drain and trap; ½in pipe.

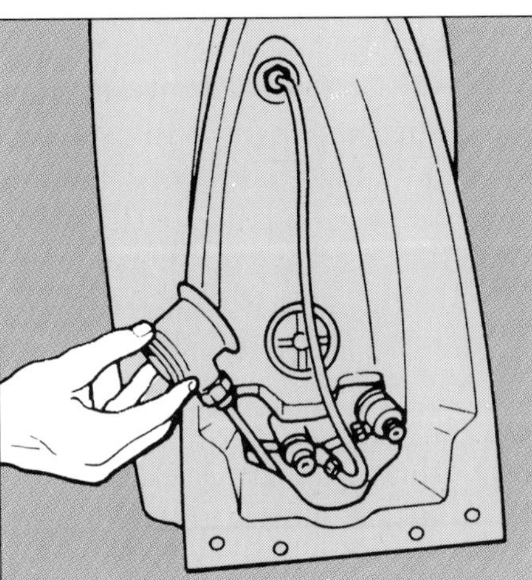

Fitting the pop-up waste to the outlet after connecting the spray supply pipe between the spray and the mixer.

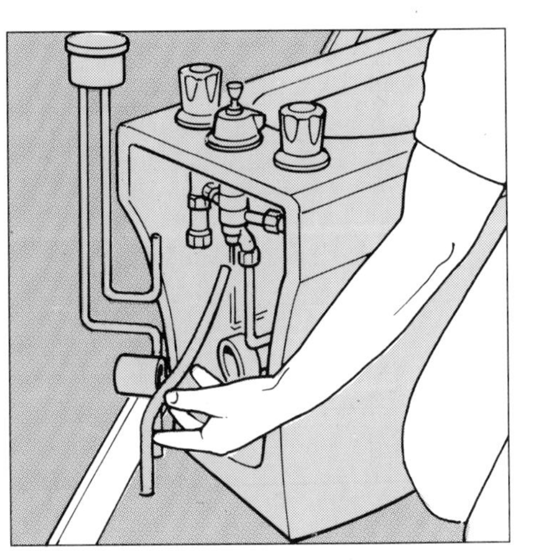

Connecting the bidet to the waste and supply pipes; currugated connectors are useful here as the bidet fits against the wall.

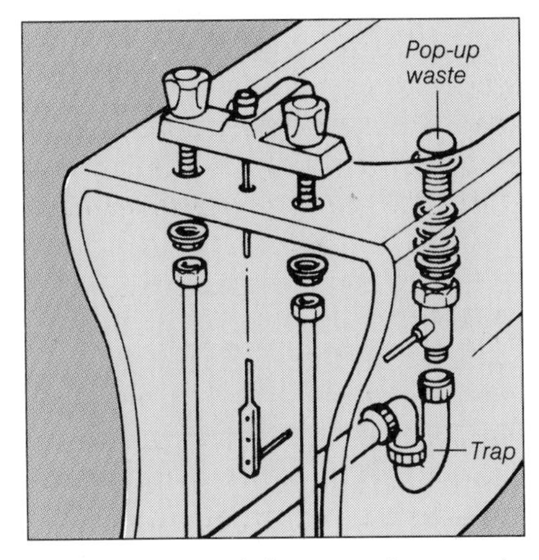

Components needed to connect an over-rim bidet: faucets; 1¼in pop-up waste and trap; ½in supply pipes.

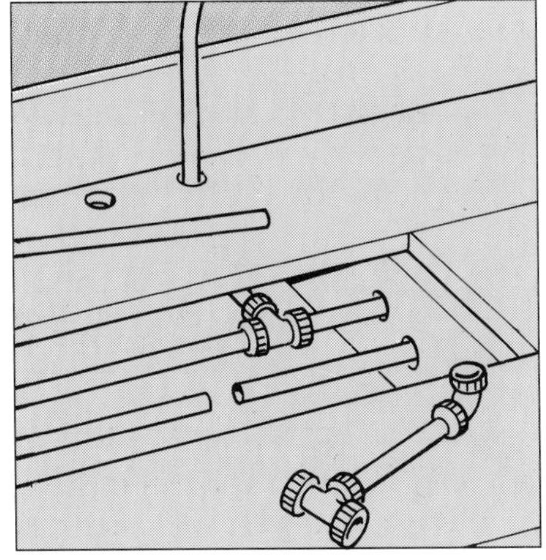

Connecting into the washbasin supply pipes; use slip-tee fittings which can be slid along the pipe after cutting.

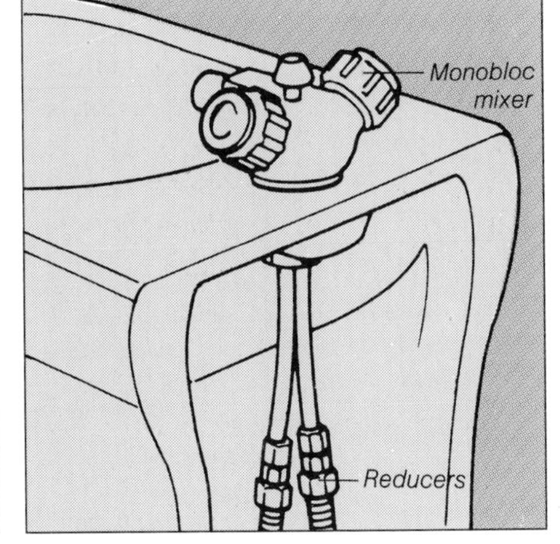

½ × ⅜in or ½ × ⅘in reducing connectors are needed to connect to the ⅜in or ⅘in tails of a monobloc mixer faucet.

Installing the bidet

With an over-rim type, assemble the faucets and waste to it as you would a normal hand basin. Since it will be set against the wall, flexible copper pipes with swivel connectors will make the connections to the ½in supply pipes much easier. Run these pipes back to the most convenient break-in points on the bathroom hot and cold faucet supplies. Note that if the cold faucets in the bathroom are supplied direct from the mains, you must have separate hot and cold faucets on the bidet and not a mixer.

The ascending spray type of bidet is somewhat more complicated to install because of its control mechanism which must be assembled to the bowl first. All the necessary parts will be supplied with the bidet and it is just a case of putting it together in accordance with the manufacturer's instructions, but always check with your local Plumbing Code before plumbing in any new appliance or fixture. Then flexible copper pipes are linked to the faucet tails as before. The cold water supply pipe must run from the base of the main cold water supply and the hot water pipe from the draw-off pipe above the hot water cylinder.

Connect up the waste pipe and screw the bowl to the floor.

THE SHOWER UNIT

There is much to be said for installing a shower – either instead of, or in addition to a bath. A shower is a very hygienic and quick way of washing, it is refreshing and what is more to shower well you will use but a fifth of the water needed for a bath. This provides real savings in the cost of heating water since you will be using less.

Where space is at a premium, a shower can be installed over an existing bath; even so, a free-standing or built-in shower cubicle takes up very little space – about 3ft square is all you really need – and it is ideal for putting in a bedroom, on a landing or in any spare corner – even under the stairs.

Installation of a shower is not at all difficult, provided that in pre-planning you have matched the arrangement to the existing plumbing system. Obtaining a good, stable pressure of water at the shower head is essential and this requirement will determine the type of fitting you use and how you arrange the pipework and connections. If this is not done properly, you may well end up with a dribble of water rather than the strong, invigorating spray you hope for. If the connections are made incorrectly, you also run the risk of water temperature fluctuations which could lead to discomfort and at worst scalding.

Types of shower unit

The simplest form of shower fitting is the old rubber hose accessory for pushing on to the bath faucets. The temperature of the spray from its rose is controlled by opening or closing both the hot and cold faucets, but it is difficult to balance them properly. Many frown on this accessory, too, since it could easily fall into the bath water, allowing the supply to become contaminated.

An improvement is the combined bath/shower mixer which is similar in appearance to a normal bath mixer faucet except that a flexible or rigid pipe runs from it up the wall to a head. To operate it, you adjust the hot and cold controls until the water flowing from the spout is the right temperature. Then you pull a small plunger to divert the water up to the head.

Unfortunately, this fitting still has the drawback of being connected to a supply that may also feed other fittings and pressure fluctuations may occur when they are used.

A better idea is the manual shower mixer which fits to the wall above the bath or in its own shower cubicle and has its own hot and cold water supply pipes. Most have two controls: one to vary the water pressure and the other to balance the hot and cold water for the desired temperature.

A thermostatic mixer will be similar in appearance and work in a similar way, but it automatically compensates for a drop of pressure in one supply pipe by reducing the flow of water from the other. In this way, the temperature set by the user is maintained, even though the overall water pressure from the head may drop. If a drastic drop in pressure from one pipe occurs, it will turn itself off completely.

Where it is impossible to connect to a cold water storage tank, you can fit an electric instantaneous shower which takes a mains supply and heats the water as it flows through, using powerful heating elements.

SHOWER SET-UPS

If a shower is to be of any use at all it must be provided with decent and constant water pressure. This need determines the way in which you connect up the pipework and also the extra equipment you may have to install. Just what you have to do will depend on the existing water supply network in your home and also on the type of home it is.

Left: Space-saving corner shower unit. Made from tough, hardwearing fiberglass, this neo-angle one-piece shower unit is designed especially for corner installation. The shower measures 38 × 38 × 721 × 4in. It has molded-in soap holders at arm level and its plain surfaces are easily cleaned and maintained. The non-sag base is sturdy and has a textured, reinforced and slip-resistant surface. Such showers can easily be fitted into existing bathrooms or bedrooms – especially where space is limited – and are often the ideal solution for home improvement schemes.

Above: Three-piece shower stall. Specially designed in three molded sections for easy installation in remodeling and new construction. Made from tough, fiberglass reinforced polyester, the shower unit comes in a wide range of colors and in two widths of 36 and 48 inches.

Screwing a shower runner to the wall after sliding the rose-holder onto the tubular runner.

Marking the pipe positions for a surface-mounted mixer after fixing the backplate to the wall.

Screwing a fixed shower rose into its holder; connect the pipework to the back of the holder.

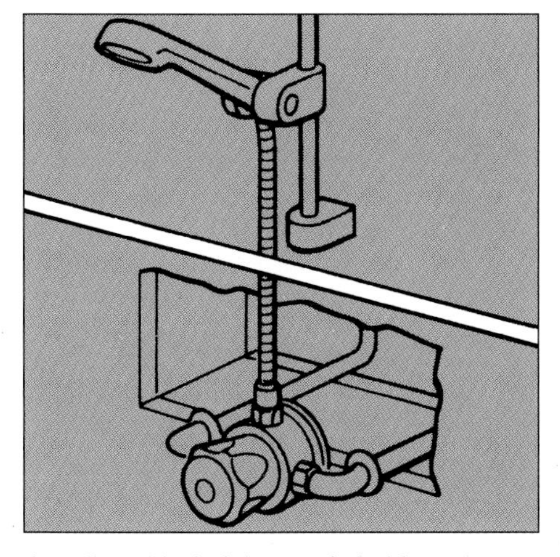

An adjustable-height rose is fed from the mixer by a reinforced flexible hose screwed onto the mixer outlet.

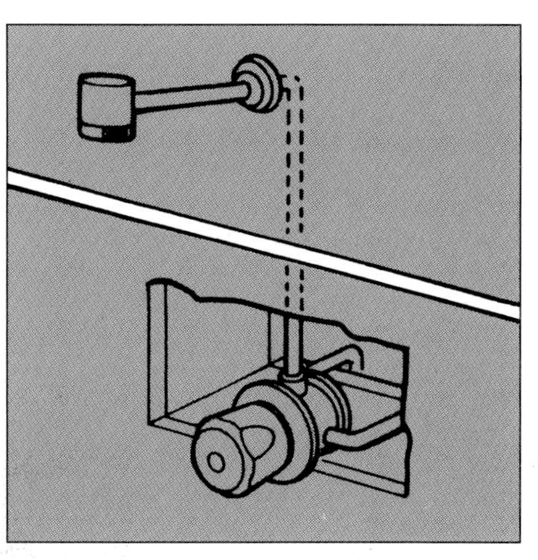

Hidden pipework is better for a fixed rose; run it up behind, or inside larger pipe within the wall.

The hot supply can be teed into a nearby hot pipe, using a ¾ × ¾ × ½in reducing tee if connecting into 22mm pipework.

FITTING A THERMOSTAT SHOWER

As with all major plumbing jobs, installing a shower, whether it is to be fitted over a bath or in its own purpose-built cubicle, requires a certain amount of pre-planning if all is to go as it should. You have to work out just where you will fit the shower head and the mixer and how you will run the pipes between the mixer and the hot and cold water supplies. Fortunately, the use of a thermostatic mixer will reduce the need to run completely new pipes from the cold water supply and the hot water cylinder. However, you should try to do this if you can.

Choosing a thermostatic mixer

There is quite a range of thermostatic mixers available and it is really a case of finding one that suits your budget and one that you like the look of. Some may be surface-mounted, which makes for easier fixing, particularly if you are mounting it over a bath; others are intended to be recessed into the wall and are more suitable for fitting into a purpose-built cubicle where this can be arranged or into a false wall specially built for the job.

Running in the pipes

Invariably the supply pipes will have to be connected to the rear of the unit regardless of whether it is surface-mounted or recessed. If the mixer is to be fixed to a false wall, this should not be too difficult to arrange since the pipes can be positioned before the wall is finally clad; the same applies if you are building a complete cubicle. However, if the wall is of masonry, it is a little more difficult.

One way is to run the pipes down the reverse face

of the wall, then pass them horizontally through it to the back of the mixer. Of course, this means that the pipes will be exposed in the adjoining room, although this may be acceptable if it is, say, a closet. On the other hand you could neatly box them in.

Setting pipes in channels

Setting the pipes in channels cut in the wall, which are then plastered over is not a good idea, since the hot pipe will expand and contract as its temperature varies and will crack the plaster. However, you may be able to get round this by running the pipes through sleeves made of pipe buried in the wall.

When laying in the pipes, work back from the mixer position to the points where you intend breaking into the existing plumbing system. That way you will only have to turn off the water for a short time while you make the final connections.

If possible, take the cold feed directly from the cold water supply and the hot feed from the draw-off pipe at the hot water cylinder immediately above the main branch for the hot faucets. If these connections would make for particularly tortuous pipe runs, you can connect into branches supplying the bathroom hot and cold faucets but *only* if you are using a thermostatic mixer. This will compensate for any drop in pressure in one pipe (caused when a faucet is turned on) by reducing the flow from the other pipe.

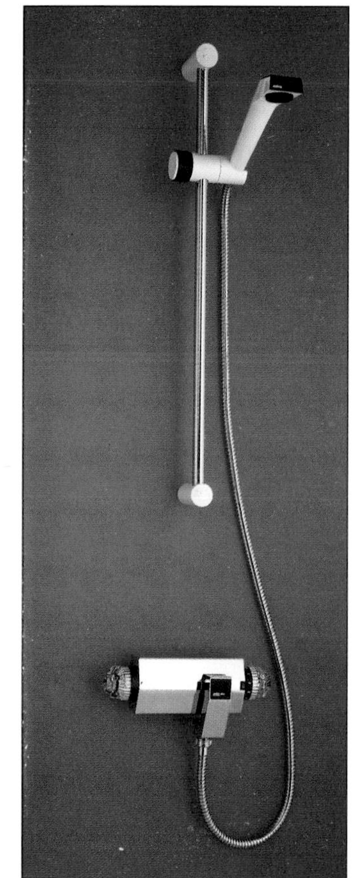

(Above) Thermostatic mixer with maximum temperature stop and dual controls for flow and temperature; can be flush or surface mounted.

(Right) Thermoscopic mixer which controls temperature to ±1°C and breaks up water scale before it can deposit; ideal in hard water areas.

Installing the mixer

The way the mixer is attached to the wall will vary from one model to another. However, surface-mounted versions will probably have a baseplate that is screwed to the wall and recessed versions will fix to the framework of the wall with screws. They have a bezel that seals to the wall to prevent water seeping behind and into the wall.

Attach the pipes to the inlets of the mixer using the compression fittings provided, making them tight.

Fitting the rose

Various types of shower heads are available: some are fixed permanently to the wall, others are part of a handset which can be slotted into a wall-mounted holder. In some cases the holder will slide up and down a vertical rod to allow height adjustment. The latter is probably the most useful.

If the head is fixed permanently it will be connected to the mixer by a rigid pipe; either concealed behind the wall or run across the surface. If the head is part of the handset, it will be connected by a flexible pipe. In each case, the connections are made with capnuts and rubber seals.

Installation is straightforward: simply screw the bracket or brackets to the wall so that the head will be at a comfortable height. Fit the head in place and connect up the supply pipe between the head and the mixer.

Water-saving devices

Shower flow control measures include fittings to be installed in standard shower heads as well as new "low-flow" shower heads designed especially to conserve water.

In both cases the flow of water is reduced to about 3 gallons per minute (compared to 5 to 8 gallons per minute with a standard head). Since most of that water is heated a considerable amount of energy, and cost, is saved, as well as water.

Mixing valves are other saving devices which allow you to mix the hot and cold water in the shower to a preset point. In this way, you save both water and energy (cost) because the temperature is guaranteed where you want it, even if you shut off the shower for a minute.

8kW unit with automatic on/off water valve permits a good flow of hot water even under very cold conditions.

High wattage electric unit with low setting for summer economy and fine temperature control.

Microchip control of temperature and flow; dial selection of flow and push-button stop/start and hot/cold.

Electric showers

If it seems too complicated to install a conventional gravity fed or pumped shower, you could install an instantaneous electric shower instead. This takes its water supply through one ½in branch pipe connected to the rising main. It also needs connecting to a 30A electrical supply controlled by a ceiling-mounted pull switch with ground fault protection.

Electric showers are housed in self-contained watertight casings so they can be installed inside the shower cubicle or over the bath. The supply pipe and cable are fed in through the back of the unit so you will need to chase the wall or run them in from the other side of the wall.

Break into the main with a T fitting and use a threaded connector to link the pipe to the shower inlet. It is essential that the switch is pull-cord operated for safety. As with any new fixture or fitting check with your local Electric and Plumbing Codes to see if it is allowed in your area – or if there are any special requirements needed before installation.

Warning

Electrical installations in bathrooms can be dangerous, and perhaps illegal, unless they strictly conform to the standards laid down in your local Building Code. So before you begin work on the project, make sure you understand thoroughly the requirements for wiring in a bathroom.

FITTING A SHOWER TRAY

Available in a variety of colors to match other bathroom fittings, shower trays can be made from enamelled cast iron or steel, acrylic plastic or ceramic. Both metal and ceramic trays are more expensive but much more hard wearing. They are also much heavier than plastic and may require two people to lift them in position. Most trays are produced in sizes ranging from 2ft 6in to 3ft square, and are designed to stand on the floor, with a surrounding apron of about 6in deep.

Because a shower tray will be close to the floor, it should be fitted with a shallow 2in seal P-trap and a 1½in waste pipe. Even with a shallow-seal trap it will be necessary to raise the tray slightly above the floor to gain the necessary clearance for the waste run. Some manufacturers supply brackets for this purpose, and others are produced with adjustable feet for leveling the tray, but if not you can set the tray on a wooden platform or on bricks.

The gap round the bottom of the tray can be filled in with tiled panels to match the cubicle and at least one should be removable to provide access for clearing blockages in the trap. You could run the trap and pipe below floor-board level but only if the pipe run is parallel to the joists. A circular waste outlet is fitted in the bottom of the tray (like a bath) but it is not usually slotted to accept an overflow.

Assembly of the trap to the tray is the same as assembling a bath waste.

After bedding the waste outlet into a ring of plumber's putty and wrapping Teflon tape round the thread, tightening the backnut.

Screwing the shallow-seal trap to the waste outlet; check that the O-ring is in place and that the trap points towards the waste pipe.

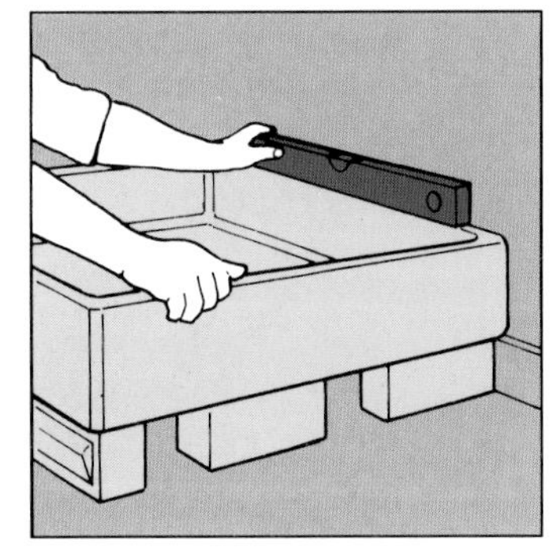
Leveling the tray on a plinth of bricks using packing pieces after connecting the waste pipe to the trap outlet.

Lowering a shower tray with its own plinth onto the adjustable feet. First locate their position and screw them to the floor.

After leveling the tray on its feet, screwing the fixing bracket to the wall; recess it into the plaster and tile over it later.

Fitting the removable side panels of the plinth after making the waste connections by pushing them up under the lip of the tray.

SHOWER CUBICLE (STALL) STYLES

The following options are available from most plumbers merchants.

Lightweight fiberglass units usually called shower wall surrounds, are the easiest to work with; they can be extremely quickly positioned and attached to framing members. Their smooth surfaces and rounded corners make them easy to keep clean.

Metal shower stalls are, perhaps, less popular these days. They are made from either tin or stainless steel. Tin is expensive and noisy because of the vibration caused by the pressure of falling water. Stainless steel is more expensive and better looking, but these surfaces do retain water stains.

Tiled-wall showers have a molded shower base which sits on a waterproof shower foundation. The ceramic tile surface can be either individualy applied tiles or pregrouted panels.

•CHECKPOINT•

Fitting a cubicle

Because a shower head is fitted high up on the wall, you must take particular care to prevent water spraying out into the room where it will damage flooring and decorations. If the shower is fitted over a bath this is easily arranged; all that is necessary is to screw a rigid shower screen to the wall so that it projects along the edge of the bath, or to install a plastic shower curtain so that it hangs inside the bath. The job is slightly more complicated if you have installed a separate shower tray, since it must be enclosed on all four sides.

There is no reason why you should not build your own cubicle, erecting wooden framed panels around it and cladding them with tiles or a similar waterproof finish, and using a shower curtain for the door. However, it is probably easier to use one of the many ready made shower kits currently available, and it would certainly be much quicker.

Cubicle format

Shower cubicle kits come in three basic forms to match the position of the tray within the room. If you have put the tray in the corner of the room so that two sides of the cubicle are already formed by the existing walls, you will need a corner kit with two upright panels. If the tray simply backs onto a wall, you can use a freestanding kit of three panels, and if the tray is in an existing alcove, all you need is a single panel kit to close it off.

These shower cubicle kits usually comprise adjustable aluminium frames (allowing different size trays to be accommodated) glazed with either patterned safety glass or, more likely, plastic acrylic sheets. One or two of the panels will incorporate sliding or bi-folding doors and you should be able to position these to give the easiest way of stepping in and out.

Securing the panels

Installation is straightforward: vertical channels are screwed to the adjoining walls, the upright panels are secured to them and the whole cubicle is held rigid by channel sections around the top and bottom which are held together at the corners by plastic blocks. A watertight seal is provided at the walls and around the tray by applying a bead of flexible caulking.

Inside the cubicle, the existing walls of the room should be tiled to prevent water seeping into the plaster and damaging the structure.

Joining the mitered corners of the bottom track by attaching the angle bracket with self-tapping screws.

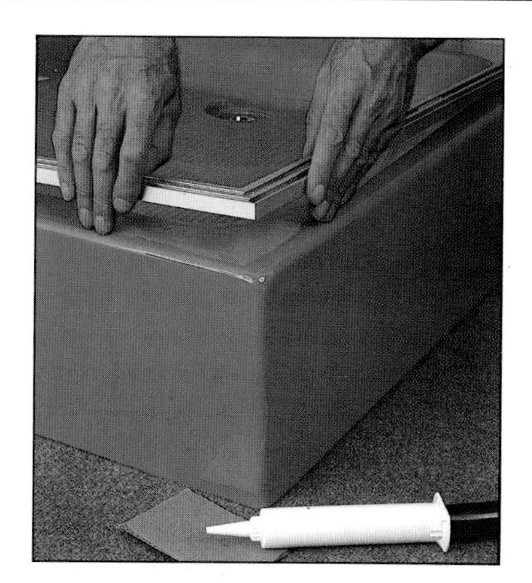

Installing the bottom track on the tray after applying a bead of caulk to the underside; ensure the angle is square.

Marking the positions for fixing the upright channel (panel spacer) to the wall; check that it is vertical.

Fitting the rollers and guide pegs to the sliding door by tapping them into the end of the door channel.

Securing the fixed panel to the panel spacer; insert the panel into the spacer and drill holes for the fixing screws.

Hanging the sliding door; position the bottom guide pegs in the bottom track and feed the top track on to the rollers.

FAULTS	CAUSES	ACTION
Faucets		
Faucet cannot be turned off fully. Constant dripping from faucet spout.	Worn faucet washer or seating.	Replace faucet washer or regrind seating.
Water flows from around the faucet spindle. Faucet handle easily turned. Juddering or knocking sound from system when faucet turned on.	Worn gland packing in old faucet; worn O-ring seals in more modern types.	Replace gland packing with wool soaked in petroleum jelly; fit new O-ring seals.
Toilets		
Clogged toilet.	Blockage in drain.	Remove blockage with plunger or closet auger.
Inadequate flush.	Faulty linkage between handle and trip lever.	Tighten setscrew on handle linkage, or replace handle.
	Tank stopper closes before it empties.	Adjust stopper guide rod or chain.
	Leak between tank and bowl.	Tighten locknuts under tank, or replace gasket.
	Clogged flush passages.	Prod obstructions from passages with wire.
Sweating tank.	Condensation.	Insulate the inside of the tank with a special liner sold at plumbing supply stores, or line with foam rubber. If the water entering the tank is often below 50°F you need a tempering valve that mixes hot and cold water.
Water flows from toilet or storage tank overflow pipe; can be heard running constantly through tank ball-valve.	Leaking ball-valve float keeping valve open.	Unscrew float and fit a new one.
	Incorrectly adjusted float arm keeping valve open.	Bend float arm so that valve closes.
	Particle of grit jammed in valve keeping it open.	Work float arm up and down to free grit; if this does not work clean valve.
	Worn ballvalve washer, preventing it closing fully.	Remove piston and replace valve washer.
	Worn ballvalve.	Replace valve.
Toilet tank flush handle needs operating several times before toilet will flush.	Worn siphon washers or flap valves in tank.	Fit new siphon washers or flat valves.
	Wear and corrosion in an old high level tank.	Replace tank.
Water flows constantly into the bowl from an old high-level toilet tank.	Worn or badly corroded "Burlington" cistern.	Replace tank.
Toilet bowl will not clear when flushed.	Blocked toilet trap.	Clear blockage with proprietary toilet plunger, or work it free with a length of flexible curtain wire.
Water level in bowl higher than normal.	Blockage in soil stack or underground drainage run.	Clear blockage with drain rods via stack clean out or access chamber.

FAULTS	CAUSES	ACTION
Juddering or knocking sound from the plumbing system when any appliance is used.	Ball-valve float in storage or toilet tank "bouncing" on ripples caused as water flows in.	Dampen float arm movement by suspending a small plastic pot from it with galvanized wire so that pot is submerged in water.
	Long unsupported pipe runs. See also second cause above.	Clip pipe run securely to walls and roof members.
Waste systems		
Waste water will not flow away from a sink, basin, bath, or other fitting.	Waste outlet blocked by materials such as soap, hair or food debris.	Clean waste outlet.
	Waste trap blocked.	Clear blockage with a sink plunger or dismantle trap and remove blockage.
Water flows from around a manhole cover.	Blockage in underground drainage run.	Clear blockage with drain rods through access chamber.
Gutter overflows.	Leaking gutter or downpipe joint.	Scoop debris from gutter and flush through; clear downpipe with a wad of rags on the end of a long rod. Remake joint.
	Sagging gutter run.	Reposition gutter brackets to ensure a constant fall towards the outlet.
Hot water systems		
No hot water from gas water heater.	The pilot light is extinguished.	Clean dust and lint from the area and relight the pilot according to the manufacturer's instructions.
	The gas inlet valve on the supply pipe has been inadvertently closed.	Open it so that the handle is parallel to the pipe. Then relight the pilot as above.
	The thermostat has been shut off or is defective.	Turn the thermostat switch to the ON position. If that does not solve the problem, replace the thermostat.
No hot water from an electric water heater.	The heater switch has been shut off.	Turn it to ON.
	The fuse has blown or the circuit breaker supplying the heater has tripped.	Restore the power.
	The upper heating element is burned out or has a calcium build-up.	Get a technician to check this out and make any necessary replacements.
Insufficient hot water or water not hot enough from an electric water heater.	The thermostat is set too low or is defective.	Call the utility company. Do not disturb the thermostat from under the insulation.
	The lower heating element has burned out.	Call a technician to replace it.
Insufficient hot water or water not hot enough from a gas water heater.	The thermostat dial is set too low.	If it is in view, adjust the thermostat upward – up to 160° for dishwasher efficiency. If the thermostat is not in view, call the utility company.
	The gas inlet valve on the supply pipe is partially closed.	Open it fully so that the handle is parallel with the pipe.

INDEX

Picture credit

Page 54: Space-saving corner shower unit by Eljer, 901 10th Street, P.O. Box 869037, Plano, Texas 75086-9037, USA
Three-piece shower stall by Lasco Bathware, 3255 East Miraloma Ave., Anaheim, CA 92806 USA